Number Theory in Mathematics Education

Perspectives and Prospects

STUDIES IN MATHEMATICAL THINKING AND LEARNING
Alan H. Schoenfeld, Series Editor

Artzt/Armour-Thomas • *Becoming a Reflective Mathematics Teacher: A Guide for Observation and Self-Assessment*
Baroody/Dowker (Eds.) • *The Development of Arithmetic Concepts and Skills: Constructing Adaptive Expertise*
Boaler • *Experiencing School Mathematics: Traditional and Reform Approaches to Teaching and Their Impact on Student Learning*
Carpenter/Fennema/Romberg (Eds.) • *Rational Numbers: An Integration of Research*
Cobb/Bauersfeld (Eds.) • *The Emergence of Mathematical Meaning: Interaction in Classroom Cultures*
Cohen • *Teachers' Professional Development and the Elementary Mathematics Classroom: Bringing Understandings to Light*
Clements/Sarama/DiBiase (Eds.) • *Engaging Young Children in Mathematics: Standards for Early Childhood Mathematics Education*
English (Ed.) • *Mathematical and Analogical Reasoning of Young Learners*
English (Ed.) • *Mathematical Reasoning: Analogies, Metaphors, and Images*
Fennema/Nelson (Eds.) • *Mathematics Teachers in Transition*
Fennema/Romberg (Eds.) • *Mathematics Classrooms That Promote Understanding*
Fernandez/Yoshida • *Lesson Study: A Japanese Approach to Improving Mathematics Teachings and Learning*
Lajoie • *Reflections on Statistics: Learning, Teaching, and Assessment in Grades K–12*
Lehrer/Chazan (Eds.) • *Designing Learning Environments for Developing Understanding of Geometry and Space*
Ma • *Knowing and Teaching Elementary Mathematics: Teachers' Understanding of Fundamental Mathematics in China and the United States*
Martin • *Mathematics Success and Failure Among African-American Youth: The Roles of Sociohistorical Context, Community Forces, School Influence, and Individual Agency*
Reed • *Word Problems: Research and Curriculum Reform*
Romberg/Fennema/Carpenter (Eds.) • *Integrating Research on the Graphical Representation of Functions*
Romberg/Carpenter/Dremock (Eds.) • *Understanding Mathematics and Science Matters*
Schliemann/Carraher/Brizuela (Eds.) • *Bringing out the Algebraic Character of Arithmetic: From Children's Ideas to Classroom Practice*
Schoenfeld (Ed.) • *Mathematical Thinking and Problem Solving*
Senk/Thompson (Eds.) • *Standards-Based School Mathematics Curricula: What Are They? What Do Students Learn?*
Sternberg/Ben-Zeev (Eds.) • *The Nature of Mathematical Thinking*
Watson • *Statistical Literacy at School: Growth and Goals*
Watson/Mason • *Mathematics as a Constructive Activity: Learners Generating Examples*
Wilcox/Lanier (Eds.) • *Using Assessment to Reshape Mathematics Teaching: A Casebook for Teachers and Teacher Educators, Curriculum and Staff Development Specialists*
Wood/Nelson/Warfield (Eds.) • *Beyond Classical Pedagogy: Teaching Elementary School Mathematics*
Zazkis/Campbell (Eds.) • *Number Theory in Mathematics Education: Perspectives and Prospects*

Number Theory in Mathematics Education

Perspectives and Prospects

Edited by

Rina Zazkis
Stephen R. Campbell
Simon Fraser University

LEA
LAWRENCE ERLBAUM ASSOCIATES, PUBLISHERS
2006 Mahwah, New Jersey London

Lawrence Erlbaum Associates, Inc., Publishers
10 Industrial Avenue
Mahwah, New Jersey 07430
www.erlbaum.com

Cover design by Tomai Maridou

Library of Congress Cataloging-in-Publication Data

Number theory in mathematics education : perspectives and prospects / edited by Rina Zazkis, Stephen R. Campbell.
p. cm. — (Studies in mathematical thinking and learning)
Includes bibliographical references and index.
ISBN 0-8058-5407-X (cloth : alk. paper)
ISBN 0-8058-5408-8 (pbk. : alk. paper)
1. Number theory—Study and teaching. I. Zazkis, Rina. II. Campbell, Stephen R. III. Series.
QA241.N8679 2006
512.7'071—dc22 2005052147
CIP

Books published by Lawrence Erlbaum Associates are printed on acid-free paper, and their bindings are chosen for strength and durability.

Printed in the United States of America
10 9 8 7 6 5 4 3 2 1

Contents

List of Contributors

Stephen R. Campbell
Simon Fraser University

David Ginat
Tel-Aviv University

José Guzmán
Cinvestav–Centro de Investigacion y de Estudios Advanzados del IPN

Carolyn Kieran
Université du Québec à Montréal

Ilana Lavy
Emek Yezreel Academic College

Roza Leikin
University of Haifa

Peter Liljedahl
Simon Fraser University

John H. Mason
Open University

Nathalie Sinclair
Michigan State University

Jennifer Christian Smith
University of Texas at Austin

Rina Zazkis
Simon Fraser University

Preface

Until recently, little attention has been given to number theory, either in teaching and learning mathematics or in mathematics education research. This is surprising given the central role number theory has played in the history and philosophy of mathematics. All the more so, when one considers that the basic concepts of number theory provide a readily accessible venue for learners of all ages to explore mathematical patterns and principles, and thereby for teachers to cultivate fundamental and deep understandings of mathematics. This book is a result of our ongoing effort to claim a more central place for number theory in mathematics education research, alongside with traditional topics such as arithmetic and algebra, and more recently introduced topics, such as statistics.

This book is of appeal to anyone interested in keeping mathematical content central in the discussion of mathematics education and mathematics education research. It will mostly attract mathematics educators, researchers in the field of mathematics education, and instructors focusing on education of teachers. Mathematics teachers and instructors of undergraduate mathematics courses that include topics from number theory will also find it relevant to their practice.

The chapters in this book offer multiple interconnected perspectives on the largely untapped potential elementary number theory has in mathematics education; its formal and cognitive nature, its relation to arithmetic and algebra, its beauty, its accessibility, its utility and intrinsic merits, to name just a few. The chapters draw on a variety of theoretical perspectives; however, the unifying perspective that binds these studies is to lay the ground for the consolidating role number theory has with regard to related pre-k–16 topics in the study of mathematics. This role is highlighted and elaborated on in the chapter by Campbell, who also addresses the more general challenge as to how to resolve and reconcile formal and intuitive understandings of mathematics.

The introductory chapter by Zazkis and Campbell provides a brief summary of the educational research related to number theory both as a content domain and as a context for investigating other issues of interest, such as proof or transition to algebra. Whereas some chapters deal with the understanding of particular concepts related to number structure and number theory (chaps. 6, 9, 10, and 11), others focus on the role these concepts could play in the development of mathematical reasoning (chaps. 2 and 3) or the interrelationship between technique and theory (chap. 8).

Furthermore, the chapters address appreciating the aesthetics of number structure (chap. 4), its suitability with regard to making connections leading to aha! insights and reaching toward learner's affective domain (chap. 7), reexamining previously constructed knowledge from a novel perspective (chap. 11), and striving for a smooth transition and extension from pattern recognition (chap. 5). Moreover, there is a significant emphasis on teacher education and the development of mathematical knowledge in teachers (chaps. 5, 6, 7, and 11) and on the use of computers and calculators as pedagogical tools that aid students' discovery on one hand, and provide an opportunity to observe concept development in students on the other hand (chaps. 4, 8, 9, and 10).

The chapters in this book highlight number theory related topics as a stepping stone toward generalization and algebraic formalism and as a means of providing intuitively grounded meanings of variables, functions, and proofs. Our purpose is to promote explication and critical dialogue regarding these issues within the broader context of the international mathematics education community. These studies comprise a variety of pedagogical and research orientations by an international group of researchers that, collectively, make a compelling case for the relevance and importance of number theory in mathematics education in both pre-k–16 and mathematics teacher education.

Rina Zazkis
Stephen Campbell

Chapter 1

Number Theory in Mathematics Education Research: Perspectives and Prospects

Rina Zazkis
Stephen R. Campbell
Simon Fraser University

> *The higher arithmetic presents us with an inexhaustible store of interesting truths—of truths too, which are not isolated, but stand in a close internal connection, and between which, as our knowledge increases, we are continually discovering new and sometimes wholly unexpected ties. A great part of its theories derives an additional charm from the peculiarity that important propositions, with the impress of simplicity upon them, are often easily discoverable by induction, and yet are of so profound a character that we cannot find their demonstration till after many vain attempts; and even then, when we do succeed, it is often by some tedious and artificial process, while the simpler methods may long remained concealed.*
>
> —C. F. Gauss (1847)[1]

What makes this book interesting and important is the detailed attention it gives to a subject area exemplifying the essence of mathematics, and one that has yet to receive sufficient attention in mathematics education. A case can be made that greater emphasis be given to number theory in mathematics education from a variety of perspectives: its formal mathematical nature,

[1]Preface to Eisenstein's *Mathematiche, Abhandlungen* (Berlin, 1847), [H. J. S. Smith], also see: H. Eves, *Mathematical Circles Adieu,* Boston: Weber and Schmidt.

particularly with respect to the additive and multiplicative structure of the integers and whole numbers; its beauty and mystique; its practical utility (or perceived lack thereof); its importance in the history of mathematics; and perhaps foremost from a pedagogical perspective, its pure simplicity and accessibility. And yet, as Gauss and so many others have long appreciated, the most trivial truths in the higher arithmetic, that is, number theory, reside side by side with many of its most profound and perennial mysteries. This volume of research expands on all of these perspectives.

Since the crash and burn of the "new math" initiatives of the 1960s, it has become customary in mathematics education to argue for teaching the practical utility of mathematics through "real-life" applications. This shift in emphasis from formalization to contextualization has experienced problems of its own. Whereas the new math curriculum left many students with deeply engrained impressions that "math is what you do in math class," the progressivist agenda has left others wondering "where's the math?"

We recognize the "practicality" of number theory in more advanced areas of application, such as cryptology and computer science, and that number theory does not readily fit progressivist preconceptions of "everyday life." Knowledge of number theory will not help in calculating taxes or balancing checkbooks. On the other hand, elementary number theory offers a natural and relatively benign algebraic formalization of integer arithmetic. Significantly, learners engaging in the subject content of elementary number theory readily encounter the very essence of mathematics. Such encounters may engender lifelong fascination rather than aversion. Accordingly, we view the utility of number theory from a different perspective—its utility for teaching and learning *mathematics*.

Topics from number theory, such as factors, divisors, multiples, and congruences provide natural avenues for developing and solidifying mathematical thinking, for developing enriched appreciation of numerical structure, especially with respect to identifying and recognizing patterns, formulating and testing conjectures, understanding principles and proofs, and justifying the truth of theorems in disciplined and reasoned ways. Considering the importance of this branch of mathematics in the history and philosophy of the discipline, the slim amount of wide-ranging and coordinated mathematics education research in this area is rather surprising—and considering the rich possibilities for research in this area, all the more so.

As the "queen" of mathematics—which is how Gauss referred to it—number theory has yet to establish a place of its own in mathematics education research. In this chapter, we discuss what this research might be and the role number theory could play in mathematics education research, not only as a "queen," but as "queen and servant." We start with portraying the presence of number theory in several official documents. Then we provide an overview of recent research and elaborate on possible future directions.

NUMBER THEORY IN "OFFICIAL" REFERENCE LITERATURE

Everyone has some familiarity with numbers. The National Council of Teachers of Mathematics (NCTM, 2000) "Principles and Standards for School Mathematics" (Principles & Standards) emphasizes that historically, numbers have been a cornerstone of the mathematics curriculum. One might risk political incorrectness and reasonably argue that numbers are in fact *the* cornerstone of the mathematics curriculum. Much emphasis has been placed in k–12 education on number operations, number sense, number systems, and even on number concepts; however, it is now becoming more widely recognized that beyond proficiency with operations, beyond familiarity with numbers, topics from elementary number theory engender and cultivate deep and fundamental understandings of mathematics, especially in areas such as problem solving and reasoning, formulating conjectures, testing generalizations, providing justifications, and proving theorems. Notably, number theory provides "a *concrete* [italics added] setting for strengthening algebraic and proof-building skills" (CBMS, 2001, p. 126).

Commenting on an NCTM (1981) commission report entitled "Guidelines for the Preparation of Teachers of Mathematics," Ball (1988) asked: "For primary grade teachers, why no course in number theory?" (p. 43). Evidently, interest in number theory has been slow to take hold in mathematics education. In the "Curriculum and Evaluation Standards for School Mathematics" (NCTM, 1989) number theory is featured only in Standard 6 for grades 5–8. The suggestion is that "the mathematics curriculum should include the study of number systems and number theory so that students can develop and apply number theory concepts (e.g. primes, factors and multiples) in real world and mathematical problem situations" (p. 91). It is further observed that "number theory offers many rich opportunities for explorations that are interesting, enjoyable and useful" (p. 91). The "Principles & Standards" recommend further attention to number theory. First, they echo the recommendation from the former document in the Number and Operations Standard for Grades 6–8, presenting the expectation that students in these grades should "use factors, multiples, prime factorization and relatively prime numbers to solve problems" (p. 214). They add the expectation that in grades 9–12 students should "use number-theory arguments to justify relationships involving whole numbers" (p. 290). If the "Principles & Standards" are any indication of curriculum trends, then we would say that there is an expectation for a stronger presence of number theory in the beginning of the new millennium.

The Conference Board of the Mathematical Sciences (CBMS, 2001) reported that "The Mathematical Education of Teachers" (MET) provides recommendations on what preservice education of mathematics teachers should include. Recognizing the special nature of the mathematical knowledge needed for teaching, these recommendations include:

- Coursework for prospective middle grades teachers should lead them to ... understand and be able to explain fundamental ideas of number theory as they apply to middle school mathematics (p. 28).
- To be well prepared to teach high school curricula, mathematics teachers need ... understanding of the ways that basic ideas of number theory and algebraic structure underlie rules for operations and expressions, equations and inequalities (p. 40).

The report further elaborates by claiming that middle school teachers

> should experience conjecturing and justifying conjectures about even and odd numbers and about prime and composite numbers. They should have a good grasp of the Prime factorization theorem and how it extends to algebra learning. The difficulty of finding the greatest common factor of two numbers can lead students to an appreciation of the efficiency of the Euclidean Algorithm. (p. 30)

Zooming out on the specifics, we acknowledge in this recommendation that number theory topics are considered an asset in gaining experience for justifying conjectures, are considered in the connection to algebra and also as an avenue for the appreciation of mathematical efficiency. We return to these themes in our discussion below of current research in these areas and of some promising prospects for further research.

A recent National Research Council (NRC) report "Adding it up" (Kilpatrick, Swafford, & Findell, 2001) has been charged with synthesizing the rich and diverse research on pre-k–8 mathematical learning and defined its explicit focus on the notion of a number. Research on whole numbers in mathematics education has traditionally focused on number operations, decimal representations, and the like. This traditional research orientation toward the number is reflected in the report, as it devotes a significant amount of space and discussion to these topics. However, mathematical learning in pre-k–8 related to the notion of number is not limited to these topics. It would be reasonable to expect that research on conceptions of number in pre-k–8, that obviously includes conceptions of number in Grades 5–8, would devote some attention to the concepts of number theory. Yet, in devoting much space and time to multiplication and division, the report does not mention concepts related to the multiplicative structure of whole numbers, such as divisibility. There is barely a mention of basic concepts from elementary number theory such as the concept of prime number, or of prime factor. There is no mention whatsoever of prime decomposition/factorization, composite numbers, or of congruence, at least in a number theoretical sense of that latter term. It seems germane to note that a complete absence from any explicit consideration of elementary concepts of number theory was also evidenced in Hiebert's (1999) review of relationships between research and the NCTM Standards. All in all, this may come as no surprise. Unlike the Standards and MET, which are works that can be seen as "agenda setting documents," the NRC's "Adding it up" is primarily a document that attempted, within limits, to review and synthesize available research. The Hiebert article, more of an illustrative review than a comprehensive one, admittedly took a prescriptive stance regarding the role of research in shaping the NCTM standards. As Kilpatrick (2001) pointed out, the NRC report, "Adding it up," also had a prescriptive component, but one that as we have seen, made no mention of number theory, and barely or nary a mention of its basic concepts. This is surprising in that, despite the pragmatic considerations involved in limiting the focus of such a work, Kilpatrick (2001) noted: (1) the main focus of the "Adding it up" report was on the topic of number in k–8; (2) number is at the heart of the school curriculum in the early grades; (3) the learning of number and related concepts is the most researched topic in the school mathematics curriculum; and (4) the learning of number and related concepts leads to the study of algebra and is strongly connected with other mathematical domains (p. 106).

To be included in the NRC "Adding it up" report, research studies had to be deemed relevant, sound, generalizable, and convergent (Kilpatrick, 2001, p. 109). A number of research studies in number theory may have conformed well to the first two criteria, but meeting the last two criteria, especially the last, requires that such studies be part of a "collection of studies ... in the sense that the findings stood up across different groups of students and teachers, were obtained using different data-gathering methods, and fit well within a larger network of evidence that made good common and theoretical sense" (Kilpatrick, 2001, p. 109). Unfortunately, much like preservice teachers' understandings of mathematics in general, research in teaching and learning number theory has been sparse and disconnected. This does not imply that research in this area has been nonexistent, but it does point to why research in this area is not represented in the "Adding it up" report.

Taken together, these studies reflect, in our opinion, an unfortunate reality—research related to or underlying elementary number theory in mathematics education has been slim to a degree of insignificance. If similar reviews are to be conducted in the next decade, will the situation be different?

NUMBER THEORY AND MATHEMATICS EDUCATION RESEARCH: LOOKING BACK AND LOOKING FORWARD

Number theory received some attention in the "new math" era (e.g., Barnett, 1961; Peck, 1961; SMSG, 1960a, 1960b, 1962). There was a minor flurry of educational interest in number theory with the advent of the microcomputer, as topics from number theory lent themselves well to programming (e.g., Anderson, 1982; Jackson, 1990; Owens, 1990). Research is continuing to flourish in this area (e.g., Abramovich, 2000; Abramovich & Brantlinger, 2004; Sinclair, Zazkis, & Liljedahl, 2003).

In the past few decades, although most often few and far between, there have been some relatively isolated studies conducted on topics ranging from the study of patterns, as a precursor to or under the rubric of algebra or pre-algebra (e.g., Healy & Hoyles, 2000; Orton & Orton, 1999) , to the role of number theory in reasoning, proof, and problem solving (King, 1973; Lester & Mau, 1993; Zazkis, 1999b). Some of these studies along with others are treated in more detail later, when we turn our focus to ways in which the queen of mathematics can, and often does, play the role of servant in mathematics education research.

There has been, at least since Ball (1988), an increasingly strong interest in number theory as a means of teaching and learning mathematics in the professional development of teachers (e.g., Campbell & Zazkis, 2002). In contrast to the relative dearth of mathematics education research in the area of number theory per se, however, there has been an ongoing vibrant treatment of the utility of number theory in teaching mathematics in the professional literature, as a quick glance at *Mathematics Teacher* (e.g., Fischer, 1986; Johnson, 2001; Kappel, 1976; Lefton, 1991; Prielipp, 1973), *Arithmetic Teacher* (e.g., Bezuszka, 1985; Esty, 1991; Lappan & Winter, 1980; Rogers, 1970), *School Science and Mathematics* (e.g., Prielipp, 1970; Woods & Flowers, 1976), and *Mathematics and Computer Education* (e.g., Anderson, 1982; Joyner & Haggard, 1990), well illustrate.

In looking back on number theory in mathematics education, we would be remiss not to mention the ongoing and enthusiastic attention given to number theory in the popular press under the rubric of recreational mathematics, notably through mathematicians like John Conway and Richard Guy (1997), and, of course, the philosopher Martin Gardner for many years, through his famous *Scientific American* column, "Mathematical Recreations."

We observed a couple of years ago that mathematics education research related to number theory is slim, especially when considering the central role that number theory has played in the development of mathematics (Campbell & Zazkis, 2002). Evidently, the statement is still true, although this volume is among the efforts to change the picture. "Related to number theory" can be interpreted in several ways: (1) There are studies concerned with learning and understanding the number theory related concepts, such as prime number or multiples; (2) There are studies concerned with more general topics, such as generalization and proof, that utilize topics from number theory as a context; (3) Within the latter, there are studies focusing on the affective domain of the learner and using number theory as a means to access this domain. In what follows, we provide a synopsis on what recent research has accomplished, what doors it opened, and what avenues are still waiting to be explored.

Number Theory as a Content Focus

The main novelty that research related to number theory brought to the field of mathematics education is in exploration of students' understanding of underlying mathematical concepts. We recall that the main criticism of anonymous "reviewer B" for our first research report submitted to the *Journal for Research in Mathematics Education* in 1994 (Zazkis & Campbell, 1996a) was that the study was not sufficiently grounded in the existing research literature. This reviewer's concern was followed with a helpful suggestion to review related literature early in the article. Reasonable as it sounds—even as novice researchers, we knew the "rules of the game"—the suggestion appeared to be problematic. Extensive literature searches uncovered reports on successful strategies for teaching number theory concepts in the middle grades, but no research on understanding these concepts. Apparently, we managed to satisfy the anonymous "reviewer B" by citing studies that used number theory as a context (see next section) and claiming that we attempt to complement these studies with investigation in which "number-theory concepts themselves are the primary focus of investigation, rather than a means to another end" (p. 540). It is notable that today researchers will not face a similar problem. There is definitely the beginning of a foundation on which to build a research agenda.

Divisibility was the main topic of our initial research (Zazkis & Campbell, 1996a). This study naturally led to several follow-up investigations, focusing on students' understanding of concepts closely related to divisibility; division algorithm and division with remainder (Campbell, 2002), even and odd numbers (Zazkis, 1998), factors and divisors (Zazkis, 2000), multiples and the least common multiple (Brown, Thomas, & Tolias, 2002), prime numbers and prime decomposition (Zazkis & Campbell, 1996b; Zazkis & Liljedahl, 2004). Several studies in this book extend the investigation on learning and understanding of particular concepts: Lavy describes a construction of the greatest common divisor concept by middle school students working in a computerized setting; Smith reaches out to more advanced concepts, focusing her investigation on preservice secondary teachers' learning about congruence relation of integers and Diophantine equations. Through the investigation of patterns, Zazkis and Liljedahl attend to divisibility, multiples, and division with remainder. The concepts of divisibility, with a particular emphasis on divisibility by 9, is further explored by Kieran and Guzman in their study of technique employed by students in problem solving. Leikin provides a perspective on understanding of prime numbers and Sieve of Eratostenes through the experience of one elementary school teacher. Liljedahl describes how students are coming to terms with divisibility and division with remainder while exploring properties of pentominoes placed on a 100-chart.

Although several pages can be easily filled when situating research on understanding number theory related concepts in the relevant literature, this is still just a beginning. Lots of concepts are left unexplored, or explored insufficiently, as well as the connections that students construct or students hold among the related concepts. For example, how do learners understand the Eucledian algorithm? What connection do they draw among different ways to determine the greatest common factor of two numbers? What are learners' understandings of relatively prime numbers? How is this understanding supported or obscured by their understanding of primes?

Furthermore, the majority of studies mentioned earlier have focused on preservice teachers. The research of Ginat, Lavi, and Kieran and Guzman (this volume) are notable exceptions. The studies of preservice teachers clearly helped in identifying common difficulties and possible avenues to overcoming those. Research on understanding number theory concepts by students, at all grade levels, is long overdue. The importance of this research reaches beyond the emphasis on specific concepts and properties. It provides a perspective for the researcher on learner's understanding of number systems and number relationships in general. Understanding relationships should be the heart of mathematics education, especially in the era of referring calculations to machines.

Number Theory as a Context

Number theory is a powerful context for conducting research in different areas. This popular choice of context by researchers can be explained following Mason (this volume), who suggests that exposure to number theory by identifying the structural sense in numbers and number sequences may provide a "paradigmatic exposure to mathematical thinking." Mason presents tasks that invite learners to generate examples of their own and through which learners develop a sense of generality in the number structure. Through these examples, learners engage in activities of generalization, recognizing invariance, and moving from recognition of patterns to awareness of underlying structure.

Conjecturing and Proving. To provide "paradigmatic exposure to mathematical thinking," the context of number theory was utilized in studies on generalization (Edwards & Zazkis, 2002) and problem solving (Lester & Mau, 1993). However, the most extensive area of research that utilized number theory as a context relates to mathematical proof and what can be considered as preformal proof territory, issues such as justification, reasoning and conjecturing.

Martin and Harel (1989) used notions of divisibility in research on preservice teachers' understanding of mathematical proof. Edwards and Zazkis (2002) used a task related to recognizing common factors and relatively prime numbers in order to explore the ways in which preservice teachers form, accept, or reject conjectures. Healy and Hoyles (2000) investigated "proof conceptions in algebra," but a majority of the statements presented in their study to be proven by participants in their research related to concepts of number theory, such as divisibility and factorization. Selden and Selden (2003) examined the ability of undergraduate students to determine the validity of an argument, that is, whether the presented argument proves a theorem. Again, the theorems they considered were from elementary number theory. Furthermore, Gholamazad, Liljedahl, and Zazkis (2003) designed a framework for analysis of short proofs written by students, by examining their proofs on the tasks of closure, that is, tasks such as "odd numbers are closed under multiplications" or "multiples of 5 are closed under addition."

Considering number theory beyond concepts and topics traditionally introduced in middle school, Rowland (2002) used theorems in number theory, such as Wilson's theorem and Chinese reminder theorem, to emphasize the generic features of proofs without relying on standard mathematical formalism. Leron (1985) adapted the Euclidian proof of the theorem for the infinity of primes to illustrate a constructive approach to indirect proofs. Harel (2002) considered theorems and tasks in number theory to investigate students' proof schemes, specifically their proof schemes related to mathematical induction. The common feature in all these studies is the selection of examples: Statements related to number theoretic properties are a popular choice among researchers in their studies and discussions related to mathematical proof.

It has been suggested that students' engagement with number theory may serve as a springboard toward appreciation and ability to construct a formal mathematical proof. For example, CBMS' report on "The mathematical education of teachers" (2001) suggested that number theory is "a concrete setting for strengthening algebraic and proof-building skills" (p. 126) and that "Many number theory topics can be avenues that lead to practice with conjecturing and proving simple theorems about numbers, thus disproving for teachers the idea that proof is only relevant in geometry" (p.

107). However, not only for teachers, but also for the majority of students, the idea of proof remains strongly connected to geometry and at times to the "two columns" way of communicating the proof. The idea of introducing learners to a formal proof via number theoretical statements awaits implementation, and the pros and cons of such implementation await detailed investigation. Will this route of initial exposure to proofs create better proof writers? Will it create a deeper understanding of what a proof entails and how to establish its validity? Maybe the next decade of research will illuminate mathematics educators with respect to these questions.

Algebra. It has been argued that elementary number theory may serve as a bridge between arithmetic and algebra. Mason (this volume) suggests that, "the common practice of detecting patterns in sequences of numbers is barely the beginning of an adventure in using natural powers to make mathematical sense and to turn arithmetic competence into algebraic awareness." Campbell (this volume) takes a stand that "number theory is already well situated between arithmetic and algebra, insofar as it is often thought of as both a generalization of whole number arithmetic and as a specialized 'subset' of algebra."

However, the relationship between arithmetic and algebra is more likely to be described as a "gap" rather than as a "bridge." The views presented by Campbell and Mason lay a theoretical foundation for curriculum development and a longitudinal empirical study: In what way can learner's engagement with number theory influence their understanding of algebra? More specifically, can it support their understanding of algebraic concepts such as functions and variables, or of algebraic notation?

On a more advanced level of mathematical sophistication, Smith (this volume) considers her study on understanding congruence as "revisiting algebra." This leads to two possible avenues of further investigation. One avenue is pursuing investigation into number theory beyond elementary topics. The growing body of research on learning undergraduate mathematics and recent recommendations for the development and implementation of "capstone" courses for prospective high school teachers (CBMS, 2001)—courses that help connect university mathematics courses with school mathematics curriculum—provide a favorable environment for this line of research.

Another avenue could focus on the issue of "revisiting." It has been argued that revisiting concepts in an unfamiliar setting can serve dual purpose. On one hand, it can serve as a fruitful pedagogical strategy for reinforcing concepts that are being revisited; Zazkis (1999a) provided a variety of examples of how this strategy can be implemented. On the other hand, such "revisiting" can serve as a tool for researchers in exploring students' understanding of these concepts. For example, researchers have shown in what way students' work with representing numbers in different bases can illuminate their understanding of the decimal numeration system (Zazkis & Khoury, 1994).

Therefore, it is reasonable to ask—and design research in an attempt to answer—in what way the work with number theory can shed light on students' understanding of "prerequisite" concepts and operations, such as multiplication and division, or division with remainder. In a way, the issue is not only with "bridging toward algebra" but also with reinforcing arithmetic. This can be rephrased using terminology introduced by Ma (1999): In what way can engagement with number theory contribute to the profound understanding of fundamental mathematics?

Furthermore, "revisiting" is also an opportunity for building connections. The MET report (CBMS, 2001) suggested that for high school teachers, "assignments in a number theory course might ask for the use of unique factorization and the Eucledian algorithm to justify familiar procedures for finding common multiples and common divisors of integers and polynomials" (p. 40). This suggestion can be seen as a research question: What connections do undergraduate students in general and prospective teachers in particular have or build between integers and polynomials as related to factorization? How can these connections be established or reinforced in an instructional setting?

Content or Context: A Blurred Distinction

In the previous sections, we discussed research related to learners' understanding of concepts in number theory, that is, the "content," and research related to notions of proof or learning algebra, in which examples from number theory were used as "context." However, having made the distinction, we suggest this distinction is often blurred and is a matter of perspective or focus.

For example, Leikin's study (this volume) can be interpreted as using the context prime numbers and Sieve of Eratosthenes to explore a teacher's learning as she is teaching a class of elementary school students. However, it can also be seen as an inquiry on a teacher's understanding of the content of prime numbers, where her teaching of elementary students serves as a context, as a window to allow this exploration. Lavi's study (this volume) can be seen as using the context of a number theory problem of creating an "n-star" to investigate students' interaction in a computerized environment. But it also can be seen as an inquiry into students' construction of number theory related concepts, where the computer microworld is a context for this investigation.

Furthermore, utilizing a number theoretical context in this volume, Ginat's study focuses on algorithmic design by computer science students.

By considering a variety of solutions to the presented tasks, he demonstrates how students' awareness and utilizing of number theory concepts and relations could lead to more elegant and efficient algorithmic solutions. Kieran and Guzman (this volume) investigate the relationship between theories students develop and techniques they employ. A task they present to students can be approached with trial and error, but design of a generalizable solution technique for the task invites consideration of number theory related concepts. In both of these studies, number theory concepts and related patterns are not crucial for a solution, but they are essential for a desirable or elegant solution. As such, students' choice of solution paths provides a window to their understanding of divisibility, multiples, or other related concepts. Taking this perspective, the studies can be seen as an investigation into understanding of specific mathematical concepts. Alternatively, they can be seen as investigation of strategy or technique, where number theory serves as means to that end.

This interplay between content and context mirrors the dual role of number theory as a queen and as a servant. Needless to say that to date, the latter has been more broadly recognized and utilized in mathematics education research.

Number Theory as Reaching Toward Learner's Affective Domain

Although engagement in number theory is certainly a cognitive endeavor, it is also an affective endeavor. That is, there is an engagement in the topic that touches the senses. In recent years, there has been an increased interest in the affective domain (McLeod, 1992) in an effort to explain why learners who possessed the cognitive resources to succeed at mathematical tasks still failed (Di Martino & Zan, 2001). This interest has been focused primarily on the more stable dimensions of learners' beliefs and attitudes about mathematics and their ability to do mathematics (Eynde, De Corte, & Verschaffel, 2001). However, there has also been an emerging interest in the less stable, more fleeting dimension of learners' emotions (DeBellis & Goldin, 1993, 1999).

In this volume, Liljedahl examines the experiences of preservice teachers as they engage in mathematical exploration. By attending to numbers a pentominoe shape covers on a 100-chart, participants made and articulated their discoveries and gained significant insight into the elementary number theory concepts that link division with remainder to multiplication with addition, and link changes in remainder to changes in the dividend. Although some of the discoveries are rather self-evident from a mathematical perspective, Liljedahl suggests that the power of learners' chains of discovery is in the transformative effect that these experiences had on these students' beliefs and attitudes toward mathematics.

Sinclair (this volume) reaches toward the affective domain of a learner through an aesthetic engagement. In fact, she argues that the aesthetic plays a role in student learning by showing how the aesthetic is intertwined with the cognitive and affective domains involved in mathematical understanding, motivation, and inquiry. Sinclair also argues that elementary number theory provides particularly rich possibilities for student aesthetic engagement and supports this argument with excerpts from several research studies.

Although the importance of affect in learning is acknowledged, the question of how to reach this domain in a subject area conservatively regarded as "pure reason" remains open. We believe, supported by personal experience and research studies such as those of Sinclair and Liljedahl, that situating learner's experiences—problems and investigations—in the content and context of number theory is one possible avenue of engaging students not only cognitively but also emotionally. Yet, the extent of this claim needs to be explored.

AN OVERVIEW OF THIS VOLUME

The following chapters present a variety of interconnected perspectives on the largely untapped potential elementary number theory has in mathematics education; its formal and cognitive nature, its relation to arithmetic and algebra, its beauty, its accessibility, its history, its utility and intrinsic merits, to name just a few. To begin with, formal, historical, and cognitive issues pertaining to number theory are considered (Campbell). Subsequently, whereas some chapters deal with the understanding of particular concepts related to number structure and number theory (Ginat, Lavi, Leikin), others focus on the role these concepts could play in the development of mathematical reasoning (Mason, Kieran, & Guzman), appreciating the aesthetics of number structure (Sinclair), its suitability with regard to making connections through "chains of discovery" (Liljedahl), reexamining previously constructed knowledge from a novel angle (Smith), to assuring a smooth transition and extension from pattern recognition (Zazkis & Liljedahl). Although each of these chapters stands on its own merit, taken together, they highlight number theory as a stepping stone toward generalization and algebraic formalism, and as a means of providing intuitively grounded meanings of variables, functions, and proofs. Our purpose here is to continue to promote explication and critical dialogue regarding these issues within the broader context of the international mathematics education community. The following chapters comprise a variety of pedagogical and research orientations by researchers in mathematics education that, collectively, make a compelling case for both the relevance and importance of number theory in mathematics education.

A FINAL COMMENT

We are about to conclude the opening chapter of the book. We believe by the book's end, with major help from the chapter authors, that collectively we will have made a compelling case for number theory in mathematics education. This includes a case for appreciating its beauty and utility in teaching and learning as well as the case for employing it as a content and context in research.

So far, research related to number theory has not "added up." We believe there is a chance for it to multiply so it will become imperative to "factor it in" in the future.

REFERENCES

Abramovich, S. (2000). Mathematical concepts as emerging tools in computing applications. *Journal of Computers in Mathematics and Science Teaching, 19*(1), 21–46.

Abramovich, S., & Brantlinger, A. (2004). Technology-motivated teaching of topics in number theory through a tool kit approach. *International Journal of Mathematical Education in Science and Technology, 35*(3), 317–333.

Anderson, S. S. (1982). Computer-assisted creativity in number theory. *Mathematics and Computer Education, 16*(2), 91–94.

Ball, D. L. (1988). Unlearning to teach mathematics. *For the Learning of Mathematics, 8*(1), 40–48.

Barnett, I. A. (1961). *Some ideas about number theory*. Reston, VA: NCTM (National Council of Teachers of Mathematics). (ERIC Document Reproduction Service No. ED175703)

Bezuszka, S. J. (1985). A test for divisibility by primes. *Arithmetic Teacher, 33*(2), 36–38.

Brown, A., Thomas, K., & Tolias, G. (2002). Conceptions of divisibility: Success and understanding. In S. R. Campbell & R. Zazkis (Eds.), *Learning and teaching number theory: Research in cognition and instruction* (pp. 41–82). Westport, CT: Ablex.

Campbell, S. R. (2002). Coming to terms with division: Preservice teachers' understanding. In S. R. Campbell & R. Zazkis (Eds.), *Learning and teaching number theory: Research in cognition and instruction* (pp. 15–40). Westport, CT: Ablex.

Campbell, S. R., & Zazkis, R. (Eds.). (2002). *Learning and teaching number theory: Research in cognition and instruction.* Westport, CT: Ablex.

CBMS (Conference Board of the Mathematical Sciences). (2001). *The mathematical education of teachers.* Providence, RI: American Mathematical Society.

Conway, J. H., & Guy, R. K. (1997). *The book of numbers*. New York: Springer-Verlag.

DeBellis, V., & Goldin, G. (1999). Aspects of affect: Mathematical intimacy, mathematical integrity. *Proceedings of 23rd Annual Conference for the Psychology of Mathematics Education* (Vol. 2, pp. 249–256).

DeBellis, V., & Goldin, G. (1993). Analysis of interaction between cognition and affect in elementary school children during problem solving. *Proceedings of 15th Annual Conference for the Psychology of Mathematics Education North America Chapter* (Vol. 2, pp. 56–62).

Di Martino, P., & Zan, R. (2001, July). Attitude toward mathematics: Some theoretical issues. *Proceedings of 25th Annual Conference for the Psychology of Mathematics Education* (Vol. 3, pp. 351–358). Utrecht, Netherlands: Utrecht University Press.

Edwards, L., & Zazkis, R. (2002). What do students do with conjectures? Preservice teachers' generalizations on a number theory task. In S. R. Campbell & R. Zazkis (Eds.), *Learning and teaching number theory: Research in cognition and instruction* (pp. 139–156). Westport, CT: Ablex.

Esty, W. W. (1991). The least common denominator. *Arithmetic Teacher, 39*(4), 6–7.

Eynde, P., De Corte, E., & Verschaffel, L. (2001, July). Problem solving in the mathematics classroom: A socio-constructivist account of the role of students, emotions. *Proceedings of 25th Annual Conference for the Psychology of Mathematics Education* (Vol. 4, pp. 25–32). Utrecht, Netherlands: Utrecht University Press.

Fischer, F. (1986). Microcomputer-assisted mathematics: Exploring number theory with a microcomputer. *Mathematics Teacher, 79*(2), 120–122

Gholamazad, S., Liljedahl, P., & Zazkis, R. (2003, July). One line proof: What can go wrong? *Proceedings of the 27th International Conference for Psychology of Mathematics Education.* Honolulu, HI.

Harel, G. (2002). The development of mathematical induction as a proof scheme: A model for DNR-based instruction. In S. R. Campbell & R. Zazkis (Eds.), *Learning and teaching number theory: Research in cognition and instruction* (pp. 185–212). Westport, CT: Ablex.

Healy, L., & Hoyles, C. (2000). A study of proof conceptions in algebra. *Journal for Research in Mathematics Education, 31*(4), 396–428.

Hiebert, J. (1999). Relationships between research and the NCTM standards. *Journal for Research in Mathematics Education, 30*(1), 3–19.

Jackson, T. H. (1990). Teaching number theory with a microcomputer. *Collegiate Microcomputer, 8*(1), 14–18.

Johnson, C. M. (2001). Functions of number theory in music. *Mathematics Teacher, 9*(8), 700–707.

Joyner, V. G., & Haggard, P. W. (1990). A classroom note on factoring and the division algorithm. *Mathematics and Computer Education, 24*(2), 154–159.

Kappel, M. W. (1976). Backtracking the Euclidean algorithm. *Mathematics Teacher, 69*(7), 598–600.

Kilpatrick, J. (2001). Understanding mathematical literacy: The contribution of research. *Educational Studies in Mathematics, 47*(1), 101–116.

Kilpatrick, J., Swafford, J., & Findell, B. (Eds.). (2001). *Adding it up: Helping children learn mathematics*. Washington, DC: National Academy Press.

King, I. L. (1973). A formative development of an elementary school unit on proof. *Journal for Research in Mathematics Education, 4*(1), 57–63.

Lappan, G., & Winter, M. J. (1980). Prime factorization. *Arithmetic Teacher, 27*(7), 24–27.

Lefton, P. (1991). Number theory and public-key cryptography. *Mathematics Teacher, 84*(1), 54–61.

Leron, U. (1985). A direct approach to indirect proofs. *Educational Studies in Mathematics, 16,* 321–325.

Lester, F. K., Jr., & Mau, S. M. (1993). Teaching mathematics via problem solving: A course for prospective elementary teachers. *For the Learning of Mathematics, 13*(2), 8–11.

Ma, L. (1999). *Knowing and teaching elementary mathematics.* Mahwah, NJ: Lawrence Erlbaum Associates.

McLeod, D. (1992). Research on the affect in mathematics education: A reconceptualization. In D. A. Grouws (Ed.), *Handbook of research on mathematics teaching and learning* (pp. 575–596). New York: Macmillan.

Martin, W. G., & Harel, G. (1989). Proof frames of preservice elementary teachers. *Journal for Research in Mathematics Education, 20,* 41–51.

National Council of Teachers of Mathematics (NCTM). (1981). *Guidelines for the preparation of teachers of mathematics.* Reston, VA: Author.

National Council of Teachers of Mathematics (NCTM). (1989). *Curriculum and evaluation standards for school mathematics.* Reston, VA: Author.

National Council of Teachers of Mathematics (NCTM). (2000). *Principles and standards for school mathematics.* Reston, VA: Author.

Orton, A., & Orton, J. (1999). Pattern and the approach to algebra. In A. Orton (Ed.), *Pattern in the teaching and learning of mathematics* (pp. 104–120). London: Cassell.

Owens, J. E. (1990). Using utility programs to explore number theory. *Journal of Computers in Mathematics and Science Teaching, 10*(1), 67–76.

Peck, L. C. (1961). *Secret codes, remainder arithmetic, and matrices.* Reston, VA: NCTM (National Council of Teachers of Mathematics). (ERIC Document Reproduction Service No. ED035548)

Prielipp, R. W. (1970). The many pearls of number theory. *School Science and Mathematics, 70*(1), 63–67.

Prielipp, R. W. (1973). The Euler phi-function and a problem of chance. *Mathematics Teacher, 66*(3), 283–286.

Rogers, H., Jr. (1970). The Euclidean algorithm as a means of simplifying fractions. *Arithmetic Teacher, 17*(9), 657–662.

Rowland, T. (2002). Generic proofs in number theory. In S. R. Campbell & R. Zazkis (Eds.), *Learning and teaching number theory: Research in cognition and instruction* (pp. 157–184). Westport, CT: Ablex.

Selden, A., & Selden, J. (2003). Validations of proofs considered as texts: Can undergraduates tell whether an argument proves a theorem? *Journal for Research in Mathematics Education, 34*(1), 4–36.

Sinclair, N., Zazkis, R., & Liljedahl, P. (2003). Number worlds: Visual and experimental access to elementary number theory concepts. *International Journal of Computers for Mathematical Learning, 8*(3), 235–263.

School Mathematics Study Group (SMSG). (1962). Study guides in mathematics: Algebra, geometry, number theory, probability, and statistics. Stanford, CA: Stanford University. (ERIC Document Reproduction Service No. ED175703)

School Mathematics Study Group (SMSG). (1960a). *Essays on number theory I.* Stanford, CA: Stanford University. (ERIC Document Reproduction Service No. ED143523)

School Mathematics Study Group (SMSG). (1960b). *Essays on number theory II.* Stanford, CA: Stanford University. (ERIC Document Reproduction Service No. ED143524)

Woods, D., & Flowers, J. D. (1976). A computer program to determine the digits of perfect numbers. *School Science and Mathematics, 76*(8), 635–638.

Zazkis, R. (1998). Odds and ends of odds and evens: An inquiry into students' understanding of even and odd numbers. *Educational Studies in Mathematics, 36*(1), 73–89.

Zazkis, R. (1999a). Challenging basic assumptions: Mathematical experiences for preservice teachers. *International Journal of Mathematics Education in Science and Technology, 30*(5), 631–650.

Zazkis, R. (1999b). Intuitive rules in number theory: Example of "the more of A, the more of B" rule implementation. *Educational Studies in Mathematics, 40*(2), 197–209.

Zazkis, R. (2000). Factors, divisors and multiples: Exploring the web of students' connections. In A. Schoenfeld, J. Kaput, & E. Dubinsky (Eds.), *Research in collegiate mathematics education* (Vol. 4, pp. 210–238). Providence, RI: American Mathematical Society.

Zazkis, R., & Campbell, S. (1996a). Divisibility and multiplicative structure of natural numbers: Preservice teachers' understanding. *Journal for Research in Mathematics Education, 27*, 540–563.

Zazkis, R., & Campbell, S. (1996b). Prime decomposition: Understanding uniqueness. *Journal of Mathematical Behavior, 15*, 207–218.

Zazkis, R. & Khoury, H. (1994). To the right of the decimal point: Preservice teachers' concepts of place value and multidigit structures. *Research in Collegiate Mathematics Education, 1*, 195–224.

Zazkis, R., & Liljedahl, P. (2004). Understanding primes: The role of representation. *Journal for Research in Mathematics Education, 35*, 164–186.

Chapter 2

Understanding Elementary Number Theory[1] in Relation to Arithmetic and Algebra

Stephen R. Campbell
Simon Fraser University

This chapter is concerned in a general sense with understanding formal mathematics and, what is closely related but not the same thing, with how mathematics can be meaningfully understood. As mathematics educators, we face an ongoing theoretical challenge as to how to resolve and reconcile formal and intuitive understandings of mathematics. A central task here is to explore ways of understanding number theory in relation to arithmetic and algebra from both formal and more intuitive points of view.

It is well known that learners are presented with many difficulties in making transitions from arithmetic to algebra. From a purely procedural view of mathematics, where strict adherence to formally defined rules of symbolic manipulation is the sole educational objective, memory would be implicated with these difficulties much more so than understanding. In a strictly procedural (or syntactic) sense, in principle there is no significant (or semantic) difference between manipulating symbols, be they numbers or letters. Such a procedural approach to mathematics, aside from stressing memory in adherence to a priori established rules, places relatively few cognitive demands on the teacher with regard to curriculum, instruction, and

[1]Elementary number theory is primarily concerned with structures and properties of the integers, and that will suffice as a working definition for our needs here. For a more detailed consideration of various facets falling under the rubric of number theory, see Campbell and Zazkis, 2002b.

assessment—that is, when compared with teaching and testing for meaningful understandings of the various concepts and relations involved.

A much wider array of cognitive and pedagogical difficulties come into play, for learners and teachers alike, when gaining and applying meaningful understandings of concepts underlying and giving rise to formal procedures are taken as important educational objectives as well. Furthermore, symbolic representations of natural numbers and operations with them are typically easier to teach and learn in concrete terms than arithmetic using rational numbers. Additional levels of abstraction presented by algebraic functions and equations are more difficult still for teachers to explain and for learners to make meaningful sense of in mathematically appropriate ways.

The foregoing considerations naturally give rise to questions regarding the role of number theory in the pre-k–16 curriculum, and why it may be important. This chapter begins with an analysis of number theory, particularly in relation to arithmetic and algebra. The chapter concludes with ways in which a phenomenological analysis of the historical and psychological epigenesis of number theory can inform traditional approaches to understanding mathematics, and thereby help to meet these challenges.

A PROBLEM CLASSIFICATION SCHEME

One way to begin understanding number theory is to identify various kinds of problems that can illustrate this area of mathematics. A problem classification scheme provides conceptual granularity in this regard. Although number theory is primarily concerned with the integers, number theory treated in k–12 and teacher education typically concerns introductory topics of number theory over the whole numbers. Topics include multiples, factors, and divisors, divisibility, divisibility rules, least common multiples (lcm), greatest common divisors (gcd), prime factorization, prime and composite numbers, prime powers, relatively prime numbers, linear sequences, and so on. Congruence relations, continued fractions, quadratic residues, Diophantine equations, and so on, can be found in undergraduate number theory courses offered by departments of mathematics in institutes of higher learning. Number theoretical topics such as these, along with the problems in number theory they give rise to, can be grouped into four overlapping problem classes; multiplicative, additive, linear, and nonlinear.

Multiplicative Problems

Multiplicative problems typically derive from the fundamental theorem of arithmetic, which asserts that every integer greater than one can be uniquely expressed, independently of order, as a product of prime numbers.

Multiplicative problems presuppose divisibility properties of integers, and are often posed in those terms. Problems that require the use, identification, or derivation of various factors, multiples, and divisors are multiplicative. The following are examples of introductory multiplicative problems from elementary number theory:

- Consider M = $3^3 \times 5^2 \times 7$. Is M divisible by 2? 3? 5? 63? (Zazkis & Campbell, 1996a)
- Find a number with exactly 13 factors (Mason, this volume).

Determining the number of factors (or divisors) of the integers provides a paradigm case for multiplicative problems in elementary number theory. For an illustration of how this problem can be implemented in the classroom, see Teppo (2002).

Additive Problems

Additive problems are typically concerned with determining sums and/or sequences of summands of integers. A classic example here is Goldbach's conjecture that every even number greater than two can be expressed as a sum of two primes. Mason (this volume), for instance, invites us to attend to the patterns in the following series of equations involving sequential sums:

$$1 + 2 = 3$$
$$4 + 5 + 6 = 7 + 8$$
$$9 + 10 + 11 + 12 = 13 + 14 + 15$$

Other well-known additive problems in elementary number theory include determining which numbers can be represented as sums of consecutive numbers, or consecutive odd numbers, or sums of squares, and so forth.

Linear Problems

Linear problems are exemplified by the Division and Euclidean Algorithms. These two algorithms are not just methods for dividing integers and determining greatest common divisors respectively, they constitute two of the most fundamental theorems of number theory. Both of these theorems are characterized by a linear structure combining multiplication and addition. Linear sequences of integers also fall under this category. Zazkis and Liljedahl (this volume) focus on repeating number patterns and their classic relation to linear sequences.

Nonlinear Problems

Nonlinear problems involve exponents of variables with values greater than one. They typically involve nonlinear equations in one or more variables such as the classic problem of determining Pythagorean triples, namely, integer solutions to the nonlinear equation: $X^2 + Y^2 = Z^2$. Fermat's generalization of this problem, and Pell's equation, $X^2 - DY^2 = 1$, provide other examples of classic nonlinear problems of elementary number theory, even though in many cases, more advanced nonintegral methods are required to solve them.

Qualifications and Limitations

Although there are many problems in elementary number theory that can readily be identified with one or another of these four classes, this classification scheme is neither disjoint nor unique. Some problems and problem classes both blur and cross the boundaries. Consider, for instance, Gauss's method for determining the sum of the first 100 nonzero integers. Prima facie, this problem is additive. A general solution to this additive problem, however, is a nonlinear function (i.e., $[n(n + 1)] / 2$). Aside from linear and nonlinear problems, there are many others with combined emphases placed on both addition and multiplication. Determining perfect numbers, in which divisors must be determined and also summed, provides a case in point. Other problems involving base representation, divisibility rules, distributivity, and congruence also place combined and deeply interrelated emphases on these two operations. Determining solutions to both linear and nonlinear equations alike are often referred to as Diophantine problems, in honor of the Hellenic mathematician, Diophantus of Alexandria (fl. ~250 A.D.).

AXIOMS OF NUMBER THEORY IN RELATION TO ARITHMETIC AND ALGEBRA

With a better sense of the various kinds of problems to be found in number theory, we can turn now to foundational axioms that can help distinguish number theory in a more formal sense from the underlying assumptions upon which k–12 arithmetic and algebra are typically based. There is a sense in which axioms can be thought of as schemas or conceptual patterns. Axioms, thus considered, are laws of thought that can be learned inductively, but insofar as they have an extension over a well-defined domain, such as the integers, they have universal application, and thus can be applied deductively. A collection of axioms defining a *ring* applies over the integers whereas a collection of axioms defining a *field* applies over the rational and real numbers. It will be helpful from a logical approach to understanding what number theory is to highlight important implications for understanding arithmetic that follow from these two formal algebraic structures.

Fields

A field is a (not necessarily numerical) set *F* with two operations, addition and multiplication, which satisfy a collection of axioms, known as field axioms. Field axioms can be separated into axioms for addition, axioms for multiplication, and the distributive axiom (i.e., distributive law) relating these two operations. Although these axioms are quite familiar, it will be helpful to enumerate them here, as they will be referred to later. Following Rudin (1976, pp. 5–6):

(A) Axioms for Addition

(A1) If $x \in F$ and $y \in F$, then their sum $x + y$ is in F.
(A2) Addition is commutative: $x + y = y + x$ for all $x, y \in F$.
(A3) Addition is associative: $(x + y) + z = x + (y + z)$ for all $x, y, z \in F$.
(A4) F contains an element 0 such that $0 + x = x$ for every $x \in F$.
(A5) To every $x \in F$ corresponds an element $-x$ such that $x + (-x) = 0$.

(M) Axioms for Multiplication

(M1) If $x \in F$ and $y \in F$, then their product xy is in F.
(M2) Multiplication is commutative: $xy = yx$ for all $x, y \in F$.
(M3) Multiplication is associative: $(xy)z = x(yz)$ for all $x, y, z \in F$.
(M4) F contains an element $1 \neq 0$ such that $1x = x$ for every $x \in F$.
(M5) To every $x \neq 0 \in F$ corresponds an element $1/x$ such that $x(1/x) = 1$.

(D) The Distributive Law

$x(y + z) = xy + xz$ holds for all $x, y, z \in F$.

The field axioms for addition (A) and multiplication (M) define the three properties of closure, commutativity, and associativity for each of these operations, whereas the distributive law (D) establishes the formal relation between them. A field also includes axioms establishing the existence of identities, that is, (A4) and (M4), and inverses, that is, (A5) and (M5), for these two operations. Uniqueness of operational identities (i.e., 0 and 1) and inverses (i.e., $-x$ and $1/x$, such that $x + (-x) = 0$, and $x(1/x) = 1$) can be logically deduced from these field axioms. The existence of unique operational inverses for all elements in the field (with, of course, the notorious exception of zero for multiplication) provides the formal basis for defining the operations of subtraction and division in terms of addition and multiplication respectively [i.e., $y - x \equiv y + (-x)$, and $y \div x \equiv y(1/x)$].

Rings

A commutative ring with (a multiplicative) identity is a (not necessarily numerical) set *R* that conforms to all of the field axioms with the crucial exception of the axiom establishing the existence of operational inverses for multiplication (M5).[2] In a commutative ring, such as the integers without multiplicative inverses, division can neither be defined nor understood in the same sense as that operation is in a field. This crucial difference regarding division in the fields of rationals and reals and the ring of integers may not come as a surprise, as there is another, more familiar way of reaching the same conclusion. It is well known that division is not closed over the integers (e.g., there is no solution for x in the equation $7 \div 5 = x$, where x is an integer). It is important to notice that this lack of closure is not due to the fact that some integers do not have multiplicative inverses. Aside from 1 and -1, there are no other multiplicative inverses in the integers. Apparent "exceptions" are a result of the divisibility property of the integers [e.g., $8 \div 4 = 2$ because $8 = 2(4)$]; they are not (partitive intuitions aside) due to multiplication by a multiplicative inverse [e.g., $8(1/4) = 2$)].

Limitations and Qualifications

The main formal difference between the real and rational number fields and the ring of integers, then, is that the integers, aside from 1 and -1, do not have multiplicative inverses. Therefore the integers are not closed over division in the same sense that the real and rational number fields are. Similarly, whole numbers do not constitute a ring because they do not have additive inverses, that is, negative numbers $-A$ such that $A + (-A) = 0$, and therefore the operation of subtraction cannot be defined in a manner that would be closed over the set. The distinctive nature and importance of whole number and integer division in providing a formal account of elementary number theory is evidenced by two of the most fundamental theo-

[2]For a ring, more generally defined, it need not be the case that for any two elements and any given operation, say a, b, and $\times$, that $a \times b = b \times a$. Nor is it necessary that there be a multiplicative identity. We are concerned here with a commutative ring with identity, that is, a ring with both additive and multiplicative identities, and in which commutativity holds.

rems of number theory; the Division Algorithm and the Euclidean Algorithm.

It is important to emphasise that I have not raised the previous considerations in some covert "neo-new math" attempt to resurrect axiomatic formalisms back into the k–12 curriculum. Rather they have been raised to emphasize, contrary to popular opinion, that there are significant differences, formally, linguistically, conceptually, and procedurally, between division with integers and whole numbers and division with rational numbers and real numbers. Evidence suggests that when these differences are conflated, as they often are, trouble follows (Campbell, 2001b, 2002a).

Indeed, the surprising and yet inevitable conclusion is that the arithmetic that we typically tend to think of in such monolithic terms that is taught in curricula throughout the world is not as monolithic as one might think. There are actually two distinct, although interrelated arithmetics involved (Campbell & Fonthal, 2000). These differences might be more evident were it not for the fact that denotational discriminations are rarely made between them. Numerals denoting whole numbers also typically designate real numbers. Moreover, the crucial terms divisor and quotient, have different meanings in whole and integer number arithmetic than they do in rational and real number arithmetic (Zazkis, 1998).

UNDERSTANDING ELEMENTARY NUMBER THEORY IN RELATION TO ARITHMETIC

Formal considerations indicate that understanding number theory in relation to and in contrast with K–12 arithmetic and algebra primarily involves questions pertaining to division. More specifically, this means, in teaching and learning number theory, that close consideration be given to understanding terms, procedures, and concepts pertaining to division with remainder (Campbell, 2002a; Zazkis, 1998), divisibility (Brown, Thomas, & Tolias, 2002; Zazkis & Campbell, 1996a), and prime decomposition (Zazkis & Campbell, 1996b).

There are reasons and evidence to suggest that when students have trouble understanding number theory, it is often because they are thinking of division with remainder (i.e., integer and whole number division) in terms of rational or real number division. Campbell (2002a) noted that whereas 10 out of 10 students with a partitive disposition toward division were unsuccessful in conducting division with remainder using a calculator, 7 out of 8 students with a quotitive disposition toward division *were* successful. Campbell suggested those observations could be accounted for by structural similarities between the Division Algorithm (i.e., for all A and $D \neq 0$, there exists unique Q and R, such that $A = QD + R$, where $0 \leq R < D$), which serves to define whole number and integer division, and the quotitive model of division.[3]

These structural similarities alone indicate the quotitive model could be a much more appropriate model for teaching and learning division with remainder than the partitive model. These considerations, of course, and can only complement the effectiveness of using the quotitive model for teaching and learning division with fractions. Partitive dispositions toward whole number division, insofar as they predispose students to divide remainders and accept nonintegral quotients as outcomes, seem at the very least to interfere with, if not completely undermine, students' understanding of number theory.

Even in the limited case of divisibility, where one might be tempted to think that it applies, the partitive model is at odds with the multiplicative structures on which the concept of divisibility is based. The very definition of divisibility, that is, for any nonzero integer D, D divides A if and only if there exists an integer Q such that A = QD, implicitly forces a quotitive interpretation of the divisor.[4] Partitive interpretations, at least insofar as they entail a whole number divisor, are more appropriately applied to teaching and learning rational and real number division, not whole number or integer division.

Understanding elementary number theory as being primarily concerned with the structure of whole numbers (and integers) helps us to understand that there are two distinct arithmetics that are typically being conceptually and procedurally conflated in pre-k–12 education. Formally, the distinction between integer arithmetic and rational number arithmetic hinges on whether the numerical domain involved constitutes a ring or a field, and as we have seen, that difference is defined by a single axiom, that is, (M5), the existence of multiplicative inverses. The conceptual link between whole number and rational number arithmetic with respect to division is nontrivial. In order to maintain a strict and clear distinction between whole numbers and rational numbers, let upper case letters represent

---

[3]Brown, Thomas, & Tolias (2002) also noted a lack of connection with multiplicative structure may be implicated in difficulties with understanding whole and integer number division.

[4]One might observe that cognitively, it is not possible to determine whether the statement A = QD is partitive or quotitive, as it can be interpreted either way. However, the quotitive model is strictly implied (i.e., the partitive model is eliminated as a viable interpretation) with the additional constraint of defining A = QD in terms of divisibility. To see this, consider first that the definition of divisibility is *not* for any nonzero quotient Q; Q divides A, if and only if there exists a divisor D such that A = QD. This fact reveals that the situation is not as simple as commutativity over multiplication initially suggests, which is what enabled the either/or interpretation of A = QD. The meanings of the divisor and quotient cannot be interchanged except in the inconsequential symmetrical case of Q = D, where Q and D are nonzero. Furthermore, in whole number and integer division, R is clearly defined as $0 \leq R < D$, and in the case of divisibility, R = 0. To interpret D partitively, where $D \neq Q$, would imply that $0 \leq Q < D$, which is clearly false.

whole number variables, and lower case letters represent rational number variables. Whole number division (division with remainder) is often expressed as A ÷ D = Q rem R (which formally speaking, is not a well formed arithmetic formula) and rational number division is expressed as a ÷ d = q. From the perspective of elementary number theory, however, it is more illuminating to compare the following:

$A = QD + R$, whole number division, where $0 \leq R < D$
$a = qd$, rational number division, where $q = ad^{-1}$ and $d^{-1} = 1/d$

The field axiom (M5) ensures that the operation of division over rational numbers can be defined by multiplication by the unique multiplicative inverse of a nonzero divisor. This comparison highlights some subtle and important similarities and differences between whole number and rational number division. Whole number division (with remainder) is revealed as a fundamental theorem of elementary number theory (Campbell, 2002a).[5] By definition, the division theorem exemplifies the fact that whole number units cannot be divided.

The crucial property of every "whole number" except zero having a multiplicative inverse in the domain of rational numbers, and the product of a "whole number" and its multiplicative inverse being unity, is exemplified in rational number division. Bearing these differences in mind, understanding number theory provides a more intuitive sense of the relation between arithmetic and algebra. This is the issue to which we now turn.

UNDERSTANDING NUMBER THEORY AS A GATEWAY TO ALGEBRA: A MORE INTUITIVE APPROACH

In some respects, number theory is already well situated between arithmetic and algebra, insofar as it is often thought of as both a generalization of whole number arithmetic and as a specialized "subset" of algebra. It has long been recognized, on the one hand, that introductory topics of number theory, such as multiples, divisors, factors, divisibility, prime factorization, the Division and Euclidean algorithms, and so forth, contribute toward a conceptual understanding of whole number arithmetic. The central theme of Standard #6 for the middle grades in the NCTM "Curriculum and Evaluation Standards for School Mathematics" (entitled "number systems and number theory") was concerned with facilitating students' understanding of the underlying structure of mathematics and of arithmetic in particular

[5]The division theorem is being used here in place of "division algorithm," which is the standard phrase used to refer to this theorem by mathematicians. As in pre-k–12 mathematics education, it is more likely to be interpreted misleadingly as the algorithm for long division. Although the division theorem is used recursively in long division, the two should not be confused one for the other.

(1989, p. 91). On the other hand, number theory can also be thought of as a subset of algebra dealing solely with the integers or whole numbers. What is more often referred to as algebra in the k–12 curriculum, however, involves working with (lower order) polynomial equations and functions over the domain of real or rational numbers. Various problems from number theory are often studied under the guise of algebra (e.g., Healy & Hoyles, 2000). Presumably, this is because problems from number theory typically involve some degree of algebraic manipulation and reasoning with variables. In these ways, it at least seems natural to think of number theory in algebraic terms, if not as a bona fide subset of algebra.

The Standard View of Number Theory in Relation to Arithmetic and Algebra

Overall, there are a wide variety of ways to think about number theory with respect to arithmetic and algebra: in terms of the relation between the concrete and the abstract, for instance; or with regard to whole and integer numbers in relation to rational and real numbers; or with regard to various coefficients, parameters, variables, and functions over these numerical domains; or with regard to reasoning and proof, and so forth—not to mention matters of instruction and pedagogy associated with the teaching and learning of all of these things. Despite its utility for helping students develop better understandings of the abstract conceptual structure of whole numbers and integers, and despite its algebraic characteristics with respect to variables and mathematical reasoning, number theory itself is often viewed by mathematics educators as being more tangential than central to the mainstream pre-k–12+ arithmetic–algebraic continuum. Indeed, this adjunct status of number theory to arithmetic and algebra appears to be so prevalent within mathematics education that it will be referred here to as *the standard view*.

One reason in support of the standard view is that number theory, as the study of the properties of the integers, constitutes a well-defined subject content area. As such, number theory can be taught and learned separately from arithmetic and algebra. Perhaps the main reason why number theory is considered tangential to the transition between arithmetic and algebra is that there is, after all, much more to both arithmetic and algebra than the domains of whole numbers or integers. Arithmetic and algebra, as taught in the k–12 curriculum, typically treat integer (and whole) numbers as subsets of real (and rational) numbers. Pedagogically, this subsumption of numerical domains is usually taken as given (e.g., Freudenthal, 1983, p. 103). Accordingly, the integers, for all intents and purposes, are taken to be *conceptually* identical to the subset of the reals known as the integers. This identification, however, is a relatively recent formal development in the history of mathematics (discussed later). An apparently related perception underlying the standard view is that it is more important for students to gain proficiency with rational numbers, and that in any case, they will gain sufficient experience with whole numbers by working in various problem-solving contexts (e.g., NCTM, October 1998, p. 206).

Even with the adjunct status of number theory in the pre-k–12 curriculum, there are good reasons for considering its role with respect to teaching and learning algebra. An ICMI discussion document (February, 2000) indicated, "... the word 'algebra' will be interpreted broadly" (p. 1), and as we have seen, there are ways in which number theory would certainly fall within that category. Indeed, the standard view of number theory constitutes an important "microcosm," or "conceptual field" (Campbell & Zazkis, 2002b) for study and analysis in addressing cognitive gaps (e.g., Herscovics & Linchevski, 1994), and the overall transition between arithmetic and algebraic thinking. The main question of concern here, however, is whether number theory can and should play a more central and less adjunct role in helping students to make that transition. In other words, just as algebra is considered the gateway to higher mathematics, is there a sense in which the study of properties of whole numbers can be considered as the gateway to algebra? With the need for more empirical research in this area, the historical development of mathematical cognition can provide some insight into the kinds of questions that need to be investigated in order to determine the extent to which number theory can contribute to the psychological development of young learners, and the extent to which it may help expedite the transition from arithmetic and serve as a gateway to algebra.

Consider, for example, the closely related set of conceptual, procedural, and semantic differences between indivisible and divisible units respectively; counting and measuring; discrete quantities and continuous magnitudes; and more generally, between whole number arithmetic and rational number arithmetic. As indicated earlier, there are good reasons and evidence to suggest that some learners are hard pressed in their understanding of differences and relations between whole number and rational number division, which in turn, are also grounded on the difference between indivisible and divisible units (Campbell, 2002a). It may be an unanticipated consequence of the standard view toward number theory that researchers, teachers, and learners are not typically aware of the manifest importance of these differences in the cognitive history of number.

A Brief Cognitive History of Number

Various aspects of the historical development of mathematical cognition may serve to inform, inspire, and enlighten teachers and students alike, but one should not thereby assume that psychological development must always be constrained or enacted in similar ways. Determining necessary (or even identifying conducive) conditions for improving the psychological development of algebraic thinking is a much more difficult question, and it may ultimately be the question on which determining the manner and extent to which number theory can serve as a gateway from arithmetic to algebra must reside. Having said that, however, there may very well be important ways in which key shifts in the historical development of Western thought actually do limit or enable the psychological development of learners' understandings of mathematics.

From a historical point of view, it is evident that the early origins of ancient Greek mathematics were predominantly concerned with patterns and properties of, and relations between, whole numbers. Different structural forms of whole numbers are evident in Pythagorean figures such as "square," "oblong," and "triangular" numbers (which I suspect were early precursors of the Platonic forms). The observation that two sequential triangular numbers form a square number provides an exemplary case of seeing the general in the particular (Mason & Pimm, 1984). Exploring why the expressions "n^2," "$n(n + 1)$," and "$n(n + 1) / 2$" provide general algebraic forms for square, oblong, and triangular numbers respectively also provides a preliminary indication of the potential role of number theory in helping learners develop more intuitively grounded meanings of variables, functions, and proofs ranging over whole numbers—thus helping them make a more meaningful conceptual transition from arithmetic to algebraic understanding. It is also evidently a part of Western cultural heritage that number theory, at least in the sense of the theoretical or conceptual (vs. the applied or procedural) study of whole numbers, preceded the development of modern algebra (Klein, 1968/1992, provides a detailed phenomenological analysis of this history). The study of Pythagorean figures provides a important clue as to the perceptual origins of numerical patterns and the conceptual origins of numerical forms, which have through the course of time come to be represented as general algebraic formulae. There is much more to the cognitive, sociocultural, and epistemological history of mathematics than this, however, and mathematics educators and researchers alike would benefit from knowing more about it than we typically do (e.g., Fauvel & van Maanen, 2000; Piaget & Garcia, 1989; Radford, 1997; Sfard, 1995). At the very least, such a history should provide an "existence proof" as to how a conceptual transition from arithmetic to algebra is possible—beginning with patterns of whole numbers. As tenuous, winding, and dark as the roads traveled over the ages might be, there are important signposts of which to take note.

Elsewhere, I have attempted to provide some insight into key metaphysical (qua cognitive) shifts involved in the developmental history of number theory that need not, for the most part, be revisited here (e.g., Campbell,

1996, 1998, 1999, 2000, 2002b). For the sake of the purpose at hand, however, I will expand on an issue that calls the standard view into question; a couple of key developments in the cognitive history of number involving the numerical unit. One crucial qualitative shift in the notion of the numerical unit was from the Pythagorean proto-atomic, spatially extended, geometrical notion of unit (Kirk & Raven, 1957/1966) to the Platonic idea of an indivisible unit with no concrete attributes whatsoever (a notion that appears to be related to Parmenides' notion of "the One," see Plato, ~388–378 BCE/1945; and for more in this regard, Campbell, 1998). A second qualitative shift of paramount importance was from the Platonic to the Aristotelian concept of a numerical unit as a divisible unit of measure (Klein, 1968/1992). To what extent might these shifts in the cognitive history of Western thought have occurred separately or independently of each other? Is there a sense in which each provides necessary or sufficient conditions for the emergence of the other? Must psychological development recapitulate historical development? More specifically, in what manner and to what extent is the cognitive history of mathematics of educational relevance?

It is helpful to recall that psychology was in the not too distant past still considered as a branch of metaphysics (Preece, 1967). If a plausible connection between the historical and psychological development of mathematical thinking is to be found, it may be evident in the historical development, or what I have been referring to as the cognitive history of mathematical thinking in Western culture. If so, then one might ask a question such as: Could the Aristotelian concept of an arithmetic unit as a continuous and divisible unit of measure have been so readily developed without the benefit of the Platonic idea of a discrete and indivisible unit? Similarly, could Platonic ideas regarding number have been developed independently of Pythagorean proto-atomic notions about the geometrical nature of number? Whether such questions can be answered though a comprehensive study of the cognitive history of mathematical thinking in the affirmative or not, simply formulating such questions can serve well to inform and guide research into the psychology of mathematics education. It will be helpful first, though, to inquire into what possible grounds there might be for relating or connecting developments in the cognitive history of Western culture with the psychological development of learners today.

Connecting History and Psychology

Phenomenology provides an important way of discerning and establishing meaningful connections between historical and psychological developments in mathematical thinking (cf. Klein, 1992, pp. 117–125; Campbell,

2001a). Phenomena, that is, the basic objects of experience, provide a natural point of departure for discerning different notions of a numerical unit. Assuming the ancients' experience of objects was not substantively different from our own, we may reasonably infer that the phenomenology, that is, the principles and structure, of their perceptual experience of objects was not substantively different from ours as well.

We learn from experience that some objects tend to present themselves as having a sense of integrity to them, such as trees, stars, and computers for instance, and there are other kinds of objects that can be more readily separated or broken apart without altering their essential nature, such as wood, water, and sand. One can break a piece of wood, for instance, into two pieces of wood, but one should usually avoid trying to break a tree into two trees.

Observation of and reflection on these qualitatively different kinds of objects provide phenomenological grounds for distinguishing among things that can be more readily or rightfully divided than others—thus providing a pedagogical basis for identifying things that cannot be broken into pieces without compromising their "integrity" (i.e., without inducing substantive changes in what they are), versus those that can.

Given there are phenomenological grounds for distinguishing between indivisible and divisible objects in a child's day-to-day lived experience, what can we possibly gain from that? A classic question of Piagetian constructivism concerns the extent to which children are capable of reflecting on and abstracting from the operational structures of their own experience, and at what point in their lives are they typically capable of doing so. More specifically of concern here is to what extent are purely conceptual abstractions—such as indivisible and divisible units, whole and rational numbers, variables, operations, functions, and reasoning over these conceptually distinct numerical domains—within the reach of young children?

It seems reasonable to hypothesize that a more phenomenologically informed instructional focus on concepts such as these from an elementary number theoretical view would prove more accessible, meaningful, and effective for children than conflating these conceptually distinct arithmetic domains and subsuming number theory into algebra over the rational numbers. This hypothesis may serve as a lynchpin with regard to whether number theory can play a more central and less tangential role in the transition between arithmetic and algebra from the early into the middle grades. Phenomenological considerations regarding the cognitive history of mathematics can provide and lend support to certain novel and innovative pedagogical practices, but ultimately determining the grade levels of their appropriateness and the degree of their success will require empirical study.

QUESTIONS FOR FURTHER RESEARCH IN MATHEMATICS EDUCATION

One approach to investigating connections between historical and psychological developments of mathematical thinking—and, more specifically, the role of number theory in developing mathematically meaningful understandings of the transition between arithmetic and algebra—is to investigate the quality of learners' understanding when curriculum and pedagogy places historical and psychological developmental trajectories in conflict. This conflict between developmental trajectories, I would suggest, is commonly the case with the standard view, which we have also seen typically involves conflating whole number and rational number arithmetic.

Children's difficulties in understanding the relations between whole numbers, fractions, and decimals are well known and have been well documented (e.g., Mack, 1995; Markovits & Sowder, 1991; Resnick et al., 1989). These kinds of difficulties can become quite deeply entrenched, manifesting themselves well into adulthood. Consequently, they appear to be difficulties that cannot be assumed to have been alleviated with subsequent exposure to algebra. Research into prospective teachers' content knowledge (a population absolutely crucial to any systemic aspirations for future improvements in mathematics education) indicates much of their understanding of basic arithmetic remains sparse, fragile, and disconnected (e.g., Ball, 1990; Campbell, 2002a; Simon, 1993). Such difficulties could be taken as an indication, if not conclusive evidence, that not clearly distinguishing between whole number arithmetic and rational number arithmetic, between whole number and rational number division, between indivisible and divisible units, and the procedural, conceptual, and semantic relations between them, presents a general problematic in understanding basic arithmetic for learners of all ages.

Conflating Whole Number and Rational Number Arithmetic

The fact that so many learners have difficulties understanding differences between various representations of whole, integer, rational, and real numbers, and the fact that so many learners experience difficulties with the transition between arithmetic and algebra may be associated with a wide variety of factors. The age, dispositions, attitudes, beliefs, socio-economic status, cultural background, gender of the learner, characteristics of teachers, parents, and peers, for instance, may all be contributing factors to one extent or another. I am suggesting, however, that a potentially crucial factor involved in making the transition between arithmetic and algebra that has apparently been overlooked, is the conflation factor; the procedural, conceptual, and discursive conflation of whole number and rational number arithmetic, and the corresponding extension of this conflation implicitly residing in the standard view toward number theory with regard to algebra. Furthermore, I am suggesting that the root of this conflation factor is grounded in a relatively recent development in the history of mathematics that has logically subsumed whole (and integer) numbers as a formal subset of rational (and real) numbers. This development appears to have motivated and encouraged some serious pedagogical mismatches between the historical, psychological, and formal development of mathematical understanding (see Campbell, 2001a).

A possible justification for living with the conflation factor, and the standard view toward number theory that appears intimately related to it, is to claim that young children are simply not developed or experienced enough to grasp the various abstract distinctions and relations to be made between whole number and rational number arithmetic. Any suggestion that instruction in rational number arithmetic be suspended, say, until the middle grades, for instance, would likely not be taken too seriously, and it is important to be clear that nothing of the sort is being suggested or implied here. Rather, from a cognitive point of view, it is more the commonly held assumption that young children are incapable of abstract thinking that is being challenged as being mistaken and misguided, both on theoretical and empirical grounds (e.g., Carraher, Schliemann, & Brizuela, 2001; Egan, 1997). Ironically, it may be the case that the cognitive difficulties in children's understanding of basic arithmetic is a result of selling short their cognitive abilities. It may also prove to be the case that their cognitive difficulties are in large part a manifestation of pedagogical mismatches between the historical and psychological development of mathematical thinking in Western culture.

Historical Development and Psychological Development

It is also important to be clear at this point that it is not being suggested or implied that the psychological development of learners should be an exact recapitulation of the historical development of mathematics, even if such a thing were possible. Such a view is clearly absurd. Rather, what is being suggested here is that a phenomenologically informed cognitive history of mathematical thinking can identify a few key moments that collectively constitute a developmentally significant sequence of critical shifts in understanding, and furthermore, that a phenomenologically informed psychological consideration of these key moments can potentially lead to significant insights regarding the nature of mathematical cognition and, thus, have significant implications for curriculum and instruction. I have offered, as a possible case in point, two key moments in the cognitive history of the development of mathematical thinking regarding the numerical unit.

Even if there are no detrimental mismatches regarding key moments in historical versus psychological development in contemporary mathematics education, how might research help to sort out the reciprocal relations and interdependencies between cognition and instruction with regard to determining the proper role of number theory in contemporary curricula? One possible framework for coordinating such a task is the notion of a conceptual field. Vergnaud (1982) has described a conceptual field as "... a set of situations, the mastering of which requires a variety of concepts, procedures and symbolic representations tightly connected to one another" (p. 36). Although much work has been done over the past 20 years in studying additive and multiplicative conceptual fields with regard to arithmetic and algebra, and although learners' understanding of some basic topics from elementary number theory have recently been studied (e.g., Zazkis & Campbell, 1996a, 1996b), a more comprehensive and systematic program of research in this area seems warranted but has yet to fully materialize—either with regard to number theory as a conceptual field unto itself, in relation to other conceptual fields, or in relation to the historical contexts from which mathematical thinking has emerged in the first place. Some recent attempts, however, have been made to help contribute to and encourage the further development of such a program of research in a variety of areas concerning cognition and instruction (e.g., Campbell, 1999; Campbell & Zazkis, 2002a).

CONCLUDING REMARKS

Although the recent National Council of Teachers of Mathematics "Principles and Standards of School Mathematics" (NCTM, 2000) provides some indication of the state and role of elementary number theory in North America, this status may vary in other parts of the world. According to the NCTM, however, the prevailing standard view of mathematics educators regarding the tangential role of number theory with regard to arithmetic and algebra implies that teachers should prioritize rational number arithmetic and that students will basically learn to discern between whole number and rational number arithmetic through the contexts in which they are used. Indeed, working with and recognizing "equivalent" numerical representations is now being stressed. Learners throughout the world, however, have notorious difficulties in understanding conceptually conflated connections between whole, fractional, and decimal representations of rational numbers. Unfortunately, for many, if not most, learners, the main commonality between arithmetic and algebraic understanding is procedural in orientation. Although learners can intuit whole number and rational number arithmetic in terms of discrete objects or continuous magnitudes,[6] respectively, k–12 students and preservice teachers alike often fail to reify the many procedural, conceptual, and linguistic distinctions that follow from those observations. Moreover, learners of all ages seem to be readily defeated when it comes to transferring the mathematical knowledge and understanding they do have beyond the specific contexts in which it was learned. On the bright side, the new "Principles and Standards" does promote the study of introductory topics of elementary number theory in the middle school curriculum, and the utility of number theory in reasoning and proof is becoming more widely recognized as well.

Historical considerations regarding the early development of mathematical cognition in Western culture suggest that number theory may have been given short shrift with regard to the importance of its role in the transition from arithmetic to algebraic thinking. It is certainly the case that the standard view tends to conflate fundamental procedural, conceptual, and linguistic distinctions between whole numbers comprised of indivisible units and rational numbers comprised of divisible units. Becoming and being phenomenologically conscious and clear about these distinctions, and by drawing on number theory to study the structure of whole numbers may provide an important psychological gateway for learners in the transition from arithmetic to algebra. As we have seen, this would likely involve relating arithmetic forms to algebraic formulae (such as visual representations of triangular numbers with "$n(n + 1) / 2$"), and by reasoning with and proving equational relationships between those arithmetic forms and algebraic formulae (such as "two consecutive triangular numbers form a square number" and $n(n + 1) / 2 + (n + 1)(n + 2) / 2 = (n + 1)^2$. In ways such as these, elementary number theory may be able to provide more intuitive and meaningful connections between whole number arithmetic and algebra.

It is quite possible that the historical tradition of Western thought could have unfolded quite differently than it has, and quite certain that it has unfolded in a much more complex manner than could ever possibly be portrayed. The question is, to what extent can historical developments in mathematical thinking, such as the shift from the Platonic idea of number (i.e., as a discrete quantity comprised of indivisible units), to the Aristotelian concept of number (i.e., as a continuous magnitude comprised of divisible

[6]Gallistel and Gelman (2000) distinguished between a discrete incrementing process, which generates countable quantities, and a continuous accumulation process, which generates continuous quantities. Citing such eminent researchers as Wynn, Dehaene, Carey, Spelke, as well as themselves in this regards, they note an emerging synthesis regarding these discrete and continuous processes, and the tensions between them, that have been "central to the historical development of mathematical thought" and "rooted in the non-verbal foundations of numerical thinking" in both nonverbal animals and humans. These processes are likely implicated in Lakoff and Núñez's (2000) four grounding metaphors of object construction and collection (viz., discrete) and measuring and motion (viz., continuous).

units), inform educational theory and practice. I have no ready answer for this question. I would suggest, however, that whatever answers one provides will inevitably have some bearing on the question as to whether or not elementary number theory constitutes either a central or tangential matter in resolving cognitive gaps and expediting the transition between arithmetic and algebraic thinking. An important question to consider in this regard is whether one can develop a purely conceptual understanding of continuous measure prior to, or perhaps even simultaneously with a conceptual understanding of discrete quantities. If so, the standard view toward the role of number theory in the curriculum might be vindicated. Whatever the case may be, I would suggest that if mathematics educators depart in their ways of teaching from the historical ontogenesis of mathematical thinking, in the sense of epic cognitive shifts such as those noted earlier, then there should be good phenomenological, psychological, and pedagogical grounds for doing so. The above considerations indicate that the opposite is true, and that more intuitive avenues for instruction are being neglected, ignored, or undervalued. In fact, the standard view of the relationship between arithmetic and algebra, and the various conflations implicit in that view are based primarily on formal developments that have emerged only recently in the historical development of mathematics.

Plato warned long ago that coming out of the cave of perceptual appearances into the dazzling and potentially blinding light of intellectual understanding too quickly could be counterproductive. The intellectual subject matter of algebra typically involves functions, equations, coefficients, and variables ranging over rational and real numbers. A clear focus on understanding properties and structures of whole numbers may provide learners with invaluable, if not necessary, scaffolding for overcoming cognitive gaps in these areas, thereby expediting the transition from the more concrete intuitive realms of arithmetic into the more lofty abstract realms of algebra. However, in order to help determine the overall effectiveness of recapitulating key developmental moments in the historical ontogenesis of mathematical thinking—in the sense of adhering to possible interdependencies of major metaphysical (qua cognitive) shifts such as those noted herein—it will be necessary to conduct longitudinal studies using pedagogical methods specifically designed to evoke those kinds of realizations.

REFERENCES

Ball, D. L. (1990). Prospective elementary and secondary teachers' understanding of division. *Journal for Research in Mathematics Education, 21*(2), 132–144.

Brown, A., Thomas, K., & Tolias, G. (2002). Conceptions of divisibility: Success and understanding. In S. R. Campbell & R. Zazkis (Eds.), *Learning and teaching number theory: Research in cognition and instruction* (pp. 41–82). Westport, CT: Ablex.

Campbell, S. R. (1996). On preservice teachers' understandings of division with remainder. In L. Puig & A. Gutiérrez (Eds.), *Proceedings of the 20th International Conference for Psychology of Mathematics Education* (Vol. 2, pp. 177–184). València, Spain: Universitat de València.

Campbell, S. R. (1998). The problem of unity and the emergence of physics, mathematics, and logic in ancient Greek thought. In L. Lentz & I. Winchester (Eds.), *Proceedings of the 1997 International Conference on the History and Philosophy of Science and Science Teaching* (pp. 143–152). Calgary, Alberta: Department of Education, University of Calgary.

Campbell, S. R. (1999). Bringing insights from research into the classroom: The case of introductory number theory. Discussion: Treating symptoms? Third Annual Conference of the Association of Mathematics Teacher Educators, Chicago, IL: AMTE. (ERIC Document ED445909). Retrieved March 24, 2002, from http://www.ceemast.csupomona.edu/amte/conference/1999/proceedings/index.html

Campbell, S. R. (2000). Zeno's paradox of plurality and proof by contradiction. *Mathematical Connections, 7*(1). Retrieved October 20, 2004, from http://www.aug.edu/dvskel/Zeno2.htm

Campbell, S. R. (2001a). Three philosophical perspectives on the relation between logic and psychology: Implications for mathematics education. *Philosophy of Mathematics Education, 14.* Retrieved October 20, 2004, from http:/ www.ex.ac.uk/~PErnest/

Campbell, S. R. (2001b). Number theory and the transition between arithmetic and algebra: Connecting history and psychology. In H. Chick, K. Stacey, J. Vincent, & J. Vincent (Eds.), *Proceedings of the 12th International Conference in Mathematical Instruction Study Group on The Future of the Teaching and Learning of Algebra* (Vol. 1, pp. 147–154). Melbourne, Australia: University of Melbourne.

Campbell, S. R. (2002a). Coming to terms with division: Preservice teachers' understanding of division. In S. R. Campbell & R. Zazkis (Eds.), *Learning and teaching number theory: Research in cognition and instruction* (pp. 15–40). Westport, CT: Greenwood.

Campbell, S. R. (2002b). What is number theory? In A. D. Cockburn & E. Nardi (Eds.), *Proceedings of the 26th International Conference for Psychology of Mathematics Education* (Vol. 1, pp. 208–212). Norwich, England: University of East Anglia.

Campbell, S. R., & Fonthal, G. (2000). Exploring whole number and rational number division within a computer-generated conceptual domain. In D. Bauder, R. Mullick, & R. Sarner (Eds.), *Proceedings of the Society for Information Technology and Teacher Education Conference* (pp. 1077–1082). San Diego, CA: AACE.

Campbell, S. R., & Zazkis, R. (Eds). (2002a). *Learning and teaching number theory: Research in cognition and instruction*. Westport, CT: Greenwood.

Campbell, S. R., & Zazkis, R. (2002b). Toward number theory as a conceptual field. In S. R. Campbell & R. Zazkis (Eds.), *Learning and teaching number theory: Research in cognition and instruction* (pp. 1–14). Westport, CT: Greenwood.

Carraher, D., Schliemann, A. D., & Brizuela, B. M. (2001). Can young children operate on unknowns? In M. v. d. Heuvel-Panhuizen (Ed.), *Proceedings of the 25th Conference of the International Group for the Psychology of Mathematics Education* (Vol. 1, pp. 130–140). Utrecht, Netherlands: Freudenthal Institute, Faculty of Mathematics and Computer Science, Utrecht University.

Egan, K. (1997). *The educated mind: How cognitive tools shape our understanding*. Chicago, IL: University of Chicago Press.

Fauvel, J., & van Maanen, J. (Eds.). (2000). *History in mathematics education*. Dordrecht, Netherlands: Kluwer.

Freudenthal, H. (1983). *Didactical phenomenology of mathematical structures*. Dordrecht, Netherlands: Reidel.

Gallistel, C. R., & Gelman R. (2002). Non-verbal numerical cognition: From reals to integers. *Trends in Cognitive Sciences, 4*(2), 59–65.

Healy, L., & Hoyles, C. (2000). A study of proof conceptions in algebra. *Journal for Research in Mathematics Education, 31*(4), 396–428.

Herscovics, N., & Linchevski, L. (1994). A cognitive gap between arithmetic and algebra. *Educational Studies in Mathematics, 27*(1), 59–78.

International Commission on Mathematical Instruction. (2000, February). *Discussion document for the twelfth ICMI study the future of the teaching and learning of algebra*. Melbourne: Author.

Kirk, G. S., & Raven, J. E. (1966). *The presocratic philosophers: A critical history with a selection of texts*. Cambridge, England: Cambridge University Press. (Original work published 1957)

Klein, J. (1992). *Greek mathematical thought and the origin of algebra*. New York: Dover. (Original work published 1968)

Lakoff, G., & Núñez, R. E. (2000). *Where mathematics comes from: How the embodied mind brings mathematics into being*. Basic Books.

Mack, N. (1995). Confounding whole-number and fraction concepts when building on informal knowledge. *Journal for Research in Mathematics Education, 26*, 422–441.

Markovits, Z., & Sowder, J. T. (1991). Students' understanding of the relationship between fractions and decimals. *Focus on Learning Problems in Mathematics, 13*(1), 3–11.

Mason, J., & Pimm, D. (1984). Generic examples: Seeing the general in the particular. *Educational Studies in Mathematics, 15*(3), 277–289.

National Council of Teaching Mathematics (NCTM). (2000). *Principles and standards for school mathematics*. Reston, VA: Author.

National Council of Teaching Mathematics (NCTM). (1998, October). *Principles and standards for school mathematics: Discussion draft*. Reston, VA: Author.

National Council of Teaching Mathematics (NCTM). (1989). *Curriculum and evaluation standards for school mathematics*. Reston, VA: Author.

Piaget, J., & Garcia R. (1989). Psychogenesis and the history of science. (H. Feider, Trans.). New York: Columbia University Press.

Plato. (1945). *The republic of Plato*. London: Oxford University Press. (Original work published ~ 388–378 B.C.E.)

Preece, W. E. (Ed.). (1967). Editor's preface. *Encyclopaedia Britannica* (Vol. 1, p. vii). Chicago: William Benton.

Radford, L. (1997). On psychology, historical epistemology and the teaching of mathematics: Towards a socio-cultural history of mathematics. *For the Learning of Mathematics, 17*(1), 26–33.

Resnick, L. B., Nesher, P., Leonard, F., Magone, M., Omanson, S., & Peled, I. (1989). Conceptual bases of arithmetic errors: The case of decimal fractions. *Journal for Research in Mathematics Education, 20*(1), 8–27.

Rudin, W. (1976). *Principles of mathematical analysis* (3rd ed.). New York: McGraw-Hill.

Sfard, A. (1995). The development of algebra: Confronting historical and psychological perspectives. *Journal of Mathematical Behavior, 14*, 15–39.

Simon, M. A. (1993). Prospective elementary teachers' knowledge of division. *Journal for Research in Mathematics Education, 24*(3), 233–254.

Teppo, A. R. (2002). Integrating content and process in classroom mathematics. In S. R. Campbell & R. Zazkis (Eds.), *Learning and teaching number theory: Research in cognition and instruction* (pp. 117–129). Westport, CT: Ablex.

Vergnaud, G. (1982). Cognitive and developmental psychology and research in mathematics education: Some theoretical and methodological issues. *For the Learning of Mathematics, 3*(2), 31–41.

Zazkis, R. (1998). Divisors and quotients: Acknowledging polysemy. *For the Learning of Mathematics, 18*(3), 27–30.

Zazkis, R., & Campbell, S. R. (1996a). Divisibility and multiplicative structure of natural numbers: Preservice teachers' understanding. *Journal for Research in Mathematics Education, 27*(5), 540–563.

Zazkis, R., & Campbell, S. R. (1996b). Prime decomposition: Understanding uniqueness. *Journal of Mathematical Behavior, 15*, 207–218.

Chapter **3**

What Makes an Example Exemplary: Pedagogical and Didactical Issues in Appreciating Multiplicative Structures

John Mason
Open University

Worked examples have always been a component of teaching mathematics, and rehearsing techniques on suites of exercises has almost always been seen as an essential part of learning mathematics. Why then has there been but limited success? The case is put forward in this chapter that in order for such examples to be useful, learners must see them as exemplifying something. If that "something" is, for example "the mystery of mathematics," "the impossibility of doing something similar myself," or "the ridiculousness of all this stuff," then learners are unlikely to make mathematical progress. If on the other hand the something is a general class of problems or other objects, and a collection of general techniques and ways of thinking, then the worked examples have served at least some of their purpose.

The transition from arithmetic to number theory is a case in point. Arithmetic is about doing calculations, but this already requires a sense of generality (e.g., number bonds to 10, methods for doing multicolumn addition, subtraction, multiplication, and division). That sense of generality is the beginning of what is referred to as *number theory,* that is, thinking in terms of collections of numbers characterized by certain properties, and learning to perceive such properties from examples. The common practice of detecting patterns in sequences of numbers is barely the beginning of an adventure in using natural powers to make mathematical sense and to turn arithmetic competence into algebraic awareness.

In order for examples to be seen as being exemplary, learners need to appreciate the generality that is being particularized. But how are they to become aware of this generality other than through examples? This apparent paradox can be sliced through by seeing both example and exemplariness as coemergently evolving awarenesses.

I suggest that when learners begin to see through a few particulars to a generality, they are beginning to formulate a theory, for the word *theory* means literally *a way of seeing*. When they can look at a problem and see it as representative of a class of problems, when they can look at an object and see it as a representative of a class of objects with common properties, then they are giving evidence of having a theory. The fact that text authors have, throughout the ages, inserted worked examples and only in some periods tried to start with general rules before illustrating their use in particular, suggests that in most generations, teachers have been aware of the difficulty of proceeding from the general to the particular (at least unless very carefully guided: see Davydov, 1990).

As George Eliot (1855) observed in an essay about Thomas Carlyle:

> It has been well said that the highest aim in education is analogous to the highest aim in mathematics, namely, not to obtain *results* but *powers*, not particular solutions, but the means by which endless solutions may be wrought. (p. 343)

Looking through a particular and seeing generality is a form of and a building block for, generating a theory. David Hilbert even formulated a principle to guide his own mathematical investigations:

> He [Hilbert] was a most concrete, intuitive mathematician who invented, and very consciously used, a principle; namely, if you want to solve a problem first strip the problem of everything that is not essential. Simplify it, specialize it as much as you can without sacrificing its core. Thus it becomes simple, as simple as can be made, without losing any of its punch, and then you solve it. The generalization is a triviality which you don't have to pay much attention to. This principle of Hilbert's proved extremely useful for him and also for others who learned it from him. Unfortunately, it has been forgotten. (Courant, 1981, p. 161)

When a learner has constructed their own theory, their own seeing of how to do a whole class of problems, then real learning is taking place. In the language of Sinclair (this book), learners participate in an aesthetic of efficiency and compactness afforded by their awareness of generality, or what the examples are exemplifying. Rowland (2002) provided further examples of techniques for supporting learners in seeing the general through the generic particular (see also Mason & Pimm, 1984).

In this chapter, I offer four case studies of classes of tasks as examples of what I call task domains. Their intention is to throw light on the exemplification paradox, and to illustrate a number of tactics that can be used to promote learner sense-making so that objects become examples of a generality. These tactics serve to enrich as well as to reveal the scope and range of learner awareness and appreciation of generality. It is argued that this is what is necessary both to be well prepared for an examination, and to understand a concept.

METHOD OF WORKING

My method of enquiry is to identify phenomena I wish to study, and to seek examples within my own experience. I then construct task exercises to offer to others to see if they recognize, or can be directed to recognize, what I find myself noticing. Through refinement and adjustment of task exercises in the light of experience and of reading relevant literature, I both extend my own awarenesses, and offer others experiences that may highlight or even awaken sensitivities and awarenesses for them. These sensitivities and awarenesses may inform their future practice. As task exercises are developed and shared, actions that exploit what is noticed for the benefit of learners are incorporated. My method does not attempt to capture or cover the experience of readers. Rather, it aims to make contact with that experience, perhaps to challenge usual interpretations, perhaps to point to some features not previously noticed but worth discerning. The data of this method are the experiences generated, the sensitivities to notice that are enhanced. If you recognize at least something of what I am talking about as a result of having worked on the task exercises, then you may be stimulated to look out for similar experiences in the future, and over time, begin to act upon what you notice. Validity in this method is personal. It depends, for you, on whether you find your actions being informed in the future. Validity for you does not lie in what I conjecture or in what I have found for myself (Mason, 2002).

CASE STUDIES

The following case studies consist of tasks that are intended to indicate (indeed to exemplify) a domain of exploration, presented in different ways to illustrate different ways in which examples can be used to illustrate and to indicate generality. Arising from these case studies will be the notion of dimensions of possible variation that is summarized in the section following the case studies.

Case Study One: Remainders

Task: One More Than. Write down a number that is one more than a multiple of 7; write down another; and another.

Experience suggests (Watson & Mason 2001, 2005) that by the third instruction, many learners are beginning to be adventurous with their numbers, and that over a period of time in which the *another and another* structure is used, learners make use of the opportunities to explore the *example space* of which they are aware. The numbers they construct shift from being *answers* to the probe, to *examples of* numbers with that property. Of course with younger learners, it would be helpful to use 10 in place of 7 so that the structure of such numbers is most evident, and then to move to less and less familiar divisors. Changing the divisor is an example of a *dimension of possible variation,* a notion that informs examplehood, and that will be elaborated in the next section.

Now a slight variation.

Task: Remainder. Write down a number that leaves a remainder of one on dividing by 7. And another, and another.

There is of course a close link between this task and the previous one, but some learners find the shift quite strange. The second task often leads people to choose a number, divide it by 7, find the remainder, and then adjust the first number, whereas the first task usually leads pretty quickly to a sense of the general. By exposing learners to a variety of responses to the task, they may become aware of an underlying structure that is captured in the particular phrasing of the first task.

These tasks are intended to promote and prompt movement from particular to general. It may take several exposures to similar tasks before learners begin to develop confidence in detecting and expressing such a generality. A learner who writes 7777 + 1 may be aware of structure whereas it is hard to know whether the response 120 is the result of structural awareness, or happenstance in finding an appropriate number.

What is it about 7778 or, more explicitly, 7777 + 1 that makes it a useful example of numbers one more than a multiple of seven? If you concentrate on or stress the repeated sevens, then you may miss the point that it can be seen as indicating any number divisible by 7. Furthermore, looked at through the focus of remainders, the remainder is what is left over after division by 7 or after removal of multiples of 7, so the 1 is seen directly as the remainder on dividing by 7. This presentation of numbers is transparent with respect to the remainder on dividing by 7, as long as you think to perceive it in this way (Zazkis & Gadowsky, 2001). But many learners do not see 7777 + 1 as a name for a number, rather, they see it as a calculation that will result in a number (Zazkis & Liljedahl, this volume).

Task: Revealing & Obscuring. Find a number for which it is easy to see the remainder of one on dividing by 7, and one that obscures the remainder.

Numbers such as 7777 + 1 and 120 serve well. In constructing examples that reveal and obscure, learner attention is directed toward the property of interest as a generality, and away from merely being satisfied with a single answer.

If the divisor was 10 rather than 7, discussion will soon reveal that it is impossible to hide the fact of a remainder of one on dividing by 10, at least when a number is expressed in base 10. Changing base, or presenting it as a product of factors might be more successful at obscuring the remainder. The structure of another and another often leads learners to increasing sophistication and awareness of a choice to be made, and this is supported by asking learners to reveal and to obscure. Their attention is extended from the particular to the class from which a particular is to be selected.

An element of competition may develop when numbers are compared, with the potential of producing useful discussion about what reveals and what obscures. The point is that as learners develop a profound understanding, the notion of *obscured* becomes increasingly problematic, impossible, or irrelevant.

Questions that prompt learners to construct not just examples, but interesting or extreme examples for themselves tend to reveal something of the scope and breadth of learners' awareness of the possibilities. To this end, Liz Bills (1996) introduced the label *particular, peculiar, general* as a reminder to prompt learners to move from particulars through examples that are peculiar or unusual in some way, in order to appreciate the general class from which those examples come, and that is exemplified by those examples.

The reason for prompting awareness of generality is that mathematics is quintessentially about seeing the general through the particular and the particular in the general, from first encounters with numerals, to the highest and most advanced of topics (Polya, 1962). The importance and significance of these moves has long been recognized:

> ... what the mathematician is seeking is Generality.... any limitation whatsoever on the generality of theorems, or of proofs, or of interpretation is abhorrent to the mathematical instinct. (Whitehead, 1911/1948, p. 57)

> To see what is general in what is particular and what is permanent in what is transitory is the aim of scientific thought. (Whitehead, 1911/1948, p. 4)

Number theory, then, can be taken as the study of properties of sets of numbers. Detecting patterns in sequences or arrays of numbers is but the first step toward appreciating number theory (i.e., numbers seen as a domain of structured properties) and serves as a paradigmatic exposure to mathematical thinking.

In the domain of numbers, particular numbers and sets or sequences of numbers are instances from the class of objects being studied, through detecting and expressing relationships, formulating relationships as properties, and reasoning with those properties. Movement between these *states of attention* is by no means trivial, and may often depend on careful teacher guidance (Mason, 2003a, reprinted in Mason, 2003b).

Case Study Two: Multiplicative Closure

The product of any two odd numbers is also odd. This well known result can be tested by recourse to examples, and might even be conjectured on the basis of experience of several examples. However, to be convinced that it is always true no matter what pair of odd numbers is chosen requires reasoning based not on knowledge about particular numbers but on the basis of a property that characterizes all odd numbers, namely, being one more than an even number, or put another way, of having a remainder of 1 when dividing by 2.

Task: Beyond Odd Numbers. What aspects or features of the statement "The product of any two odd numbers is also odd" could be changed and still the statement would remain true?

Certainly the number of odd numbers multiplied could be changed to 3, 4, and so on to any positive integer. So also the hidden two that defines odd numbers as numbers with remainder one on dividing by 2 can be varied. That divisor could also be 3 or 4, or any positive number. Indeed, as will emerge in relation to the first case study later, the divisor could be any positive number whatsoever, and by further extension, any nonzero number.

Changing odd is less clear. Thinking in terms of one more than a multiple of 2, or in terms of remainder one on dividing by 2, you could try changing the remainder. A remainder of zero reduces to thinking of multiples of 2, and if the divisor is also altered, to multiples of other positive integers. However, zero and one are the only remainders that give multiplicative closure. More adventurously, two or more remainders can be combined, considering, for example, all numbers with remainder either 0 or 1 on dividing by 7, or any other specified divisor, or all numbers whose remainders are relatively prime to the divisor for a fixed composite divisor. Sticking with 7 as the divisor, the product of any two numbers whose remainders on dividing by 7 are either 1 or 6 will also have a remainder of either 1 or 6 on dividing by 7, as will the set of numbers leaving a remainder of 0, 1, or 6 on dividing by 7.

Note the effect of being given these particular examples. How can you look through them to an encompassing generality? How can you treat them as generic? You have to consider what aspects of this example might change. In order to make sure you appreciate what is being exemplified, it might be appropriate to try some particular cases by changing the 7 to other numbers and by looking for satisfactory subsets in order to try to work out what is going on. You will of course be encountering the mathematical notion of sets of remainders that are closed under multiplication modulo the divisor.

Apart from changing the divisor, it seems you can sometimes choose a set of remainders. In particular, if you have two such sets, you can generate a new set of remainders from these, by combining the elements of the two sets of remainders appropriately. Notice this change of level from considering the structure of sets of numbers closed under multiplication, to the structure of sets of remainders that are closed under multiplication. This sort of move by shifting to a higher level of abstraction is typical of pure mathematics.

Finding sets of numbers closed under multiplication is not in itself particularly interesting, until you realize that such numbers may or may not factor into the product of numbers of the same type, which leads to primes. Whenever you can multiply two things together, there is a converse operation of factoring: When can an object be expressed as a product? Those that cannot (except in some trivial way such as by using 1 as a factor) are analogous to the prime numbers in ordinary arithmetic.

Task: Extending The Meaning of Prime. Among the set of numbers that are one more than a multiple of 3, there are some that cannot be factored into a product of two numbers of that form without using 1 as one of the factors. Such a number (excluding 1) is called *prime*.

Noting that 100 is one more than a multiple of 3, in how many different ways can you factor it as the product of numbers, each of which is also one more than a multiple of 3?

Must this phenomenon happen when you change the multiples of 3 to multiples of some other number?

The fact that 100 can be factored in more than one way using different primes shows that the fundamental theorem of arithmetic (the uniqueness of prime factoring) that holds for the positive integers, fails to hold for other multiplicatively closed sets of numbers. Unique factorization is an important property of the whole numbers because several properties of arithmetic depend on this fact. The reason that one is excluded from being prime is so as to make the statement of the fundamental theorem aesthetically concise.

Note that in order to make sense of these remarks, you have to have worked with the particular number 100, recognized that it has the property of having two different factorizations into primes within the system of numbers that are one more than a multiple of 3, and seen through the particularity of 100.

Task: Characterizing. Construct other numbers that share with 100 the property of not having unique factorization into primes within numbers of the form one more than a multiple of 3. Are there numbers with three distinct factorisations, or more? What role is played by the divisor 3?

Characterizing is a quintessentially mathematical activity. It means finding one or more properties that all such numbers have, and that distinguishes them from numbers not having that property. Even where it is difficult to achieve a characterization, it is mathematically useful to generalize the structure that gives rise to the property.

Note also that in order to appreciate the extension of the meaning of *prime* from whole numbers to other sets of numbers, it is probably useful to construct some examples for yourself, through which to appreciate the generality. At the same time, it is necessary to let go of the numbers you know familiarly as primes, because once you change systems, other numbers become prime! This flexibility of thinking is vital for effective mathematical thinking. An object that exemplifies a concept in one setting may not be an example in a different setting. For example, 10 and 25, being composite numbers ordinarily, turn out to be prime among numbers of the form one more than a multiple of 3.

Task: Multiplicatively Closed Sets of Numbers. Show that the set of those positive whole numbers that, when expressed as a product of powers of primes have 2 raised to an even power and 5 raised to a power that is a multiple of 3, is closed under multiplication. Generalize!

Note that the structure of another and another could also be used for this task, and even more explicitly, there could be a suggestion to consider dimensions of possible variation in order to promote generalization. It is by considering the general that the particular transcends the singular, the unusual, and becomes a window through which to appreciate the general.

Using primes and their powers is not the only way to form sets of multiplicatively closed numbers. Leonard Euler found a completely different approach.

Task: Euler's Sets. Show that the set of numbers that can be expressed as the sum of two squares is closed under multiplication. What dimensions of possible variation can you think of?

Starting from scratch, this task could be rather difficult. If you know about complex numbers however, you will know that the norm of a complex number is the sum of the squares of the real and imaginary parts, and that the product of the norms is the norm of the product. This may be how Euler thought of it, but the algebra has a certain appeal when you appreciate the underlying structure:

$$(a^2 + b^2)(x^2 + y^2) = (ax + by)^2 + (ay - by)^2 = (ax - by)^2 + (ay + by)^2$$

The generality needs to be interpreted in terms of products of numbers each expressible as a sum of two squares, namely that the answer can also always be expressed as the sum of two squares, and often in two different ways.

Notice how the particular special case of the sum of two squares is hard to break out of when trying to find a suitable dimension of variation, that is, some aspect or feature to vary. Attempts to increase the number of squares added together to three are doomed to failure. But Euler varied the coefficients of the summed squares and found that numbers that are the sum of a square and twice another square are also closed under multiplication. Once you have seen why this is the case, it is much easier to conjecture that perhaps the 2 of "twice a square" can be changed to other numbers. Being too bold and trying numbers of the form, say, twice a square plus three times another square do not have the desired property of closure. In the former case, Euler's reasoning can be generalized, whereas in the latter case, it cannot. Note that the coefficient "twice" can be any integer, positive, or negative!

Finally, there is the question of how Euler's forms and powers of primes fit together. A special case is Lagrange's theorem that numbers that are the sum of two squares are characterized by the property that prime divisors of the form one less than a multiple of 4 must appear to an even power. This is a difficult theorem and out of reach of most secondary students and novice teachers, but it shows how mathematical exploration can develop. Of most interest here is the role played by examples, whether of numbers with particular properties, types of tasks, or uses of core mathematical themes and thinking processes.

Case Study Three: Common Multiples and Common Divisors

Task: Common Divisors and Multiples. Find the greatest common divisor (GCD) of 84 and 90. Find their least common multiple (LCM) as well.

It is not hard to imagine a whole page of questions like this, in which the learner is asked to find common divisors and common multiples. The class of similar tasks comprises a *space of questions* (Sangwin, 2003) from which some instances or elements or examples have been drawn. As such, the questions chosen are, to a teacher or author, exemplary of the class. But are they chosen in such a way as to reveal the structure of that space to learners? Do learners also have a sense of this class, even when they have worked through the page (Watson & Mason, 2004)?

It seems that when confronted with a collection of tasks, many learners deal with each item in sequence, without trying to see a connection between them, whereas a sensible author uses systematic variation from question to question to suggest a large class of similar examples. What is being educated is the learners' awareness of what constitutes "of this type." They are becoming aware of what can be varied and what must remain the same in order to constitute a type of task. Learners who, when they meet an examination question, are not instantly aware of underlying structure or of the particular as a particular case of a general class, are at a major disadvantage. They have to solve the problem *ab initio* from their use of their own powers of mathematical thinking, whereas learners who recognize a type or structure are triggered into action using methods with which they are already familiar.

To expand learners' awareness of class rather than particular, it is useful to prompt them to make up for themselves versions that "show possible difficulties that can arise in some cases," that "illustrate what things can happen," that are peculiar or extreme in some way, or that are thought to be hard, obscure, or challenging, as suggested in previous case studies (Watson & Mason, 2005).

The topic of greatest common divisors and least common multiples is not simply about finding them in particular cases. Learners need exposure to the notion that there are multiple ways of expressing or denoting numbers, and that different representations require different techniques for considering common divisors and multiples. Expressing numbers as products of powers of primes makes LCM and GCD much more transparent (Zazkis & Gadowsky, 2001), but for this transparency, such product representations need to be seen as numbers and not just as calculations that will produce numbers (Zazkis & Liljedahl, this volume). A classic task for exposing learners to the product perspective is the following.

Task: Odd Number of Factors. Find a number with exactly 13 factors

This task starts with a particular number that is large enough to invite simplification (try three factors, try one factor!) and then regeneralization beyond the 13, with extensions concerning characterizing or describing the class of all numbers with a given odd number of factors, and finding the smallest such number. Bold explorers might consider characterizing numbers with other numbers of factors. This task is typical of the approach developed in the 1960s and 1970s in England (Banwell, Tahta, & Sounders, 1972) in that learners are invited to undertake a task in which they make choices in order to simplify on the way to recomplexifying for themselves. An alternative presentation could be to ask learners to find numbers with an odd number of factors, or even to ask learners to find the number of factors of a collection of numbers, and then to explore the consequences of sorting these into numbers with an odd number of factors and those with an even number.

Finding a number with exactly 13 factors is not a single task but an entry point into a whole domain of tasks that include finding numbers with a given number of factors, and finding how many numbers less than itself are relatively prime, among others (see Banwell et al., 1972, pp. 37, 102–103). Such a task domain includes various variants in presentation as well as in content. It is of pedagogic use only if it becomes a vehicle for learners to use their natural powers to imagine and to express, to generalize and to particularize, to conjecture and to convince, and if their attention is drawn to the fact of these powers and to ways of refining and honing them, or as Paul Goldenberg put it, to establishing mathematical habits of mind (Goldenberg, 1996). For example, although an observer might say that they detected the emergence of a theory about the structure of numbers as products of primes, and about the parity of the number of divisors, participants might be wholly unaware that that is what they were doing. Their attention is likely to be confined to trying to sort out their ideas and to justify their conjectures. Yet what they are doing is what mathematicians do. Becoming aware of the emergence and articulation of a theory can both inspire and support the development of learner's mathematical self-esteem.

Yet another version is a problem that affords access to the same awarenesses about numbers.

Task: Porters. Consider a collection of locked doors numbered 1, 2, 3, ... and a collection of porters also numbered 1, 2, 3, ... Each porter alters the locked–unlocked state of those doors marked with multiples of their particular number. Which doors are open when all of the relevant porters have done their rounds?

There are of course a number of variants and extensions. For example, what happens if only some specified porters do their rounds, and does it matter if porters do their rounds out of order?

There are also structural relations between common divisors and common multiples. Prompting learners to encounter structural relationships between GCDs and LCMs involves a shift in what learners are attending to, from calculating GCDs and LCMs, to relating them together. This can be done by inviting a return to particular examples, which anticipates that learners will then generalize for themselves, or by drawing attention to the general, and inviting learners to make choices concerning the extent of particularizing and the way of thinking about numbers as products of prime powers or as base 10 numerals. It is possible, and perhaps ultimately desirable, at least to get a taste of how it is possible to reason with the structural properties of the GCD and the LCM without using particular numbers, and perhaps, not even using a representation as a power of powers of primes, but simply reasoning from the definitions.

Task: Relations Between GCD and LCM. Particular-to-general: Multiply the GCD and LCM of 84 and 90 together, and compare with the product of 84 and 90; generalize.

General-to-particular: Find a relationship between the GCD and the LCM of two numbers.

This task hides a choice to make between thinking of numbers as products of powers of primes, and thinking of numbers only in base 10 notation.

Further developments arise when attention is directed to the fact that a pair of numbers (84 and 90) were suggested rather than, say, three or more numbers. Is the GCD of three or more numbers well defined? Can it be found by finding the GCD of pairs of numbers? What about the LCM? How are these related to the GCDs and LCMs of the numbers taken in pairs?

The next task is generic in form, switching attention from finding answers to characterizing the answers that are possible, or characterizing the initial situations that lead to a particular answer.

Task: GCD and LCM Reversal. What numbers can be the GCD or the LCM of a pair of numbers? What pairs of numbers can have 24 as their GCD? What pairs of numbers can have 72 as their LCM? Generalize!

Although stated in the particular with an implicit invitation to generalize, the task can be presented in many other forms according to the situation.

Working with a particular number such as 24 is intended to be exemplary for how it could be done in general. A learner who works on the particular but does not pay attention to how the particular example works is simply getting answers. This is in contrast to what George Polya called *specializing*, or particularizing, in order to appreciate the particular as exemplifying what happens in general. It might be useful to start with simpler numbers, but in order to reach a generality, the learner needs to construct examples that exemplify what can happen (for example, using a number such as $p^3q^4r^5$ as a particular case, where p, q, and r are distinct primes). Note the necessity of choosing a representation for numbers that is transparent with respect to factoring.

In tackling a reversed or undoing task, participants are not only developing a theory about the structure of pairs of numbers with a given LCM, they are also experiencing the use of their power to organize and characterize in a mathematical context, much as mathematician are wont to do. In other words, by being immersed in such tasks, learners are likely to develop a theory of what doing mathematics is like, a theory that would be very different from a theory developed as a result of only attending lectures and doing routine exercises (Watson & Mason, 1998).

Case Study Four: Factoring

Task: Factoring Plus or Minus. Lee Minor (1988) noticed that:

$x^2 + 5x + 6 = (x + 3)(x + 2)$ and $x^2 + 5x - 6 = (x + 6)(x - 1)$

$x^2 + 13x + 30 = (x + 10)(x + 3)$ and $x^2 + 13x - 30 = (x + 15)(x - 2)$

$x^2 + 25x + 84 = (x + 21)(x + 4)$ and $x^2 + 25x - 84 = (x + 28)(x - 3)$

$x^2 + 41x + 180 = (x + 36)(x + 5)$ and $x^2 + 41x - 180 = (x + 45)(x - 4)$

Generate the next few pairs and generalize. Justify your conjectures!

Asking learners to say what is the same and what is different about each pair may direct attention to the fact that each pair of quadratics has the same coefficients except that the sign of the constant term is altered. Yet both quadratics factor. The sequence of pairs suggests looking for patterns. Asking yourself what is invariant and what is changing as you go down the sequence is likely to reveal the sequence of whole numbers (with negative signs in the second factoring of each pair), and may lead to recognizing the presence of triangular numbers. Alternatively learners may notice patterns in the growth of the coefficients of x in the quadratics.

Anne Watson (2000) uses the term *going with the grain* to describe the detecting and following of patterns such as the whole, square, and triangular numbers and even, constant first or second differences, and so on. Detecting such patterns provides conjectures that can then be tested by expressing them algebraically and then justified by using algebraic manipulation. Going with the grain is relatively easy (although somewhat of a challenge in this case), but what is important about such tasks is not merely the exercise of powers of pattern detecting and following, but making sense of what is detected, or *going across the grain*. The metaphor develops, since wood can be split, chiseled, or planed along natural lines with the grain, but the underlying fibrous structure is revealed when you saw across the grain. So too with sequences and tables of numbers that invite generalization (Mason, 1999, 2001; Mason, Graham, & Johnston-Wilder, 2005). What is mathematically significant is the sense making, which begins with expressing generality, because this exposes the fibrous structure of the sequence. In the case of the quadratic factoring, there is an implicit general question to characterize all quadratics that have the property that changing the sign of the constant term still leaves the expression factorizable. More generally still, characterize quadratics that can be factored no matter which coefficients have their signs changed.

An alternative presentation could provide even more data, so that learners can seek a pattern grain to follow in rows and columns:

TABLE 3.1
Jointly Factored Quadratics

$x^2 + 5x + 6$ $= (x + 2)(x + 3)$ $x^2 + 5x - 6$ $= (x + 6)(x - 1)$			
$x^2 + 10x + 24$ $= (x + 6)(x + 4)$ $x^2 + 10x - 24$ $= (x + 12)(x - 2)$	$x^2 + 13x + 30$ $= (x + 3)(x + 10)$ $x^2 + 13x - 30$ $= (x + 15)(x - 2)$		
$x^2 + 17x + 60$ $= (x + 12)(x + 5)$ $x^2 + 17x - 60$ $= (x + 20)(x - 3)$	$x^2 + 20x + 96$ $= (x + 8)(x + 12)$ $x^2 + 20x - 96$ $= (x + 24)(x - 4)$	$x^2 + 25x + 84$ $= (x + 4)(x + 21)$ $x^2 + 25x + 84$ $= (x + 28)(x - 3)$	
$x^2 + 26x + 120$ $= (x + 20)(x + 6)$ $x^2 + 26x - 120$ $= (x + 30)(x - 4)$	$x^2 + 29x + 210$ $= (x + 15)(x + 14)$ $x^2 + 29x - 210$ $= (x + 35)(x - 6)$	$x^2 + 34x + 240$ $= (x + 10)(x + 24)$ $x^2 + 34x - 240$ $= (x + 40)(x - 6)$	$x^2 + 41x + 180$ $= (x + 5)(x + 36)$ $x^2 - 41x - 180$ $= (x + 45)(x - 4)$

Note: See Mason (2006).

Task: More Factoring Plus or Minus. What is going on in this table? What "with the grain" patterns can be found reading down a column or along a row? Find a general entry. Are all possible such quadratics produced?

The thrust of the task is certainly not to produce a general formula to be memorized. Rather, it is to prompt learners to practice factoring in the context of a larger goal, namely experiencing the range of possible variation that constitutes the phenomenon of quadratics that can be factored when the constant term is either positive or negative. The task structure is to offer two dimensions of possible variation corresponding to rows and columns. With any luck, it will prompt learners to try changing the sign of the coefficient of x and to discover that this does not affect whether a quadratic factors. Some learners might notice that they can insert an extra parameter, say t in the coefficient of x and a corresponding t^2 in the constant. A few learners may be tempted to try to achieve a parametrization of quadratics with a coefficient of x^2 other than one (see Kaczkowski, 2001).

This factoring task is intended to exemplify an admittedly complex version of a whole range of tasks that can be constructed by offering learners sequences or arrays of mathematical objects in which one or two parameters are being varied, with the intention that learners will detect these variations and express the generalities. The next section highlights the importance of systematic and near-simultaneous variation for learning.

DIMENSIONS OF POSSIBLE VARIATION

Ference Marton has suggested that learning means discerning details that have previously not been distinguished. Certainly discerning variation is at the core of what it means to learn a concept or a technique. The underlying mechanism for human beings and indeed for all life forms, is to discern or make distinctions by detecting variation (Marton & Booth, 1997; Marton & Trigwell, 2000; Marton & Tsui, 2004). This is the basis of our sensory input. At the cognitive level, an example of a concept is only appreciated as an example when certain features are recognized as being permitted to change, while some feature remains (relatively) invariant. These are the dimensions of variation. Since different people are aware of different dimensions, and in particular, because novices do not initially share awareness of all of the richness possible variation, Watson and Mason (2005) extended his notion to *dimensions of possible variation*. But what values can these variations take? Each dimension of possible variation has an associated *range of permissible change*, which might also not be shared between novices and experts. For example, some learners might think that to have a remainder, the divisor has to be a whole number, which is a limited range of permissible change.

The following notes draw attention to some of the ways in which dimensions of possible variation and corresponding ranges of permissible change are used in the case studies.

Case Study One

In the remainder task, the numbers 1 and 7 can both be changed, and so represent dimensions of possible variation. If learners focus on the particularity of the 1 and the 7 (or whatever numbers are used by the teacher) they may be unaware of the role of the particular as examples of something more general. In fact, the 7 is permitted to be not only any positive integer but any nonzero number (see later section on extending definitions).

There are other implicit dimensions of possible variation in the remainder tasks. You could for example, ask for a number, and indeed all numbers, with remainder 1 when dividing by 2, and with remainder 3 when dividing by 5 as well, and this is merely an indicator of a domain of exploration that culminates in the Chinese remainder theorem. A popular entry to it in Medieval puzzle books was to find a number of eggs with remainder 1 on dividing by 2, and by 3, and by 5, as in problems posed by Nicholas Chuquet (Flegg, 1985; Mason, 1996). Note, however, that the apparent invariance of the number 1 might restrict learners' awareness of possible variation when generalizing.

Another direction of development is the following task.

Task: Continued Remainders. Write down a number for which the remainder when dividing by 2 is 1, and the remainder when dividing the quotient (the whole number obtained in the previous division by ignoring the remainder) by 3 is 1, and the remainder when dividing that quotient by 4 is 1. Why must any such number be divisible by 3?

There are of course many possible dimensions of variation to open up to exploration from this task, including deciding whether *every* number divisible by 3 will have the stated property (and if not, characterizing in some other way the ones that do), as well as characterizing those sets of constraints (remainder when dividing by something is something) that permit a general conclusion about the starting number analogous to "divisible by 3."

Case Study Two

In the second case study, there is an explicit dimension of possible variation in the choice of multiples, and an implicit but also structurally complex variation in the remainder. Remainders of zero could replace remainders of 1, as could sets of remainders such as $\{1, m - 1\}$ (where multiples of m are being considered). The shift from thinking of numbers as presented in terms of products of powers of primes, to thinking in terms of forms such as the sum of a square and twice a square represents an implicit dimension of possible variation, and within that, there are possible variations with a range of permissible change for the *twice,* but not multiples of both squares. Further exploration could seek possibilities for other forms of numbers such as the sum of two cubes, sum of three or four squares, and so forth. Some of these are possible, and others are not.

Case Study Three

In the third case study, there is an implicit dimension of possible variation in the use of pairs of numbers, which can be extended to triples and beyond, as well as the explicit dimension of changing the actual numbers. The awareness that factoring numbers as products of prime powers makes all the thinking much easier but that there is an algorithm for dealing with base 10 notation is also a dimension of possible variation. Further elements of the range of permissible change could include representing numbers in terms of the Chinese remainder theorem arising from the first case study.

One way to stimulate learners to experience calculations as processes is some variant of tell an absent friend how to calculate the *GCD* and *LCM* of a pair of numbers; how to program a machine to find the *GCD* and *LCM* of a pair of numbers; or I am going to be given a pair of numbers, but I can't tell you what they are at the moment (or I have a friend who has a pair ...), so can you please tell me how to calculate their *GCD* and *LCM*.

Suddenly what seemed almost frighteningly open becomes bounded. A theory might just be possible.

Notice that I do not provide my answers or even my theories in the sense of ways of seeing. For once theories are published, pedagogic value leaks away, because the pedagogic value of a task lies in the mathematical themes that will be encountered, the mathematical powers that will be used, the mathematical techniques that will be helpful, and the mathematical concepts that may be developed or drawn on. The answer itself is rarely of abiding interest except as a check of correct reasoning. Unfortunately the *transposition didactique* (Chevallard, 1985) is a reminder that expert awareness is all too easily turned into training in behavior: What appears to its author as an intriguing task is usually transposed into a sequence of "things for learners to do" stated as unambiguously as possible, thereby removing learners' choices and reducing their opportunities to use their own mathematical powers. The transposition comes from the best of intentions to make sure everyone can get started, but when what is produced is behavior manuals (worksheet instructions), the pedagogic value has leaked away.

Case Study Four

In the fourth case study, the examples that comprise the initial data are laid out in a manner that limits the variation so that learners can detect a pattern as they locate and work with the grain. Expressing a generalization initiates working across the grain to make sense of the whole, and to raise questions about whether all such quadratics will appear in these tables, and whether changing the sign of the coefficient of x makes any difference to factorizability. All of the examples displayed have the coefficient of x^2 as an implicit value of 1. Once attention is directed to this, the possibility that this coefficient represents a dimension of possible variation extends the exploration even further.

Invariance in the Midst of Change

A ubiquitous theme throughout mathematics is the study of invariance in the midst of change: What can change and still some thing, some relationship, some property, remains the same. Many mathematical theorems can be seen as characterizing what is permitted to change and still something remains invariant, or, put another way, what remains invariant despite certain things changing. For example, the classic theorem that the sum of the interior angles of a planar triangle is 180° illustrates this well, for the smallest and apparently least significant word, "a" is the only indicator of the scope of permissible change, namely any planar triangle whatsoever, whereas "sum" specifies what is invariant, and the 180° states what that invariant actually is. Among the case studies, having an odd number of factors is an invariant property among all square numbers, and having a unique decomposition into prime factors is a property not shared by numbers one more than a multiple of 3.

Thinking in terms of dimensions of possible variation as highlighted in case study two particularly, not only provides the teacher with a reminder to be aware of the important dimensions, but can be used with learners so that they too begin to think in these terms. Furthermore, asking learners to construct objects according to various constraints and to make those objects as general or extreme or as exemplary as possible, reveals the scope of learner awareness and appreciation of the complexity of concepts and techniques. Another version of the same idea can be found in Brown and Walter (1983), who base their approach to problem posing around the question "what if ... were to change?"

One outcome of Marton's approach to learning as discernment is that appreciating invariance in the midst of change is what it means to understand and appreciate a concept. You don't really understand a concept unless you have a sense of what features of an example make it exemplary concerning the concept, that is, what aspects of an example can change and still it remains an example.

For example, to appreciate what it means to be prime, you need not only to be articulate in the use of a definition, but to appreciate some of what happens when you change from ordinary whole numbers under addition to other sets, such as numbers that are one more than a multiple of m for some m. The reason is that it is tempting to see the property of unique factorization and the notion of prime factors as glued together rather than the first as a property that may or may not hold depending on the structure under discussion. To appreciate the notions of GCD and LCM in case study three, it helps to be aware of their role as the *greatest* and the *least* respectively, which emerges from exploring the class of pairs of numbers that can have a specified number as GCD or as LCM.

Appreciating and Expressing Generality

An object can only be an example if there is a class of objects of which it is a representative, even if that class has very few members, perhaps even only one! Generalization lies at the very heart of mathematics, with justified generalization at the core of pure mathematics, and instantiated generalization at the core of applied mathematics. Gattegno pointed out that generalization consists of stressing some features and allowing them to vary or change, while ignoring other features, and that "it is a property of the mind to isolate and merge, to stress and ignore" (Gattegno, 1970, p. 136). Although there is much talk and much writing about learners constructing meaning and learners being immersed in mathematical practices, at the heart of learning mathematics lies the practice of detecting patterns and expressing generality as conjectures to be checked and justified. Expressing generality for yourself is vital in order to appreciate where algebraic expressions come from and why anyone would want to manipulate them. Once you know how algebraic expressions arise, it is much easier and attractive to engage in doing other peoples' algebra, that is, in practicing algebraic manipulations of various sorts, in order to develop facility. But if you have not produced many such expressions yourself, their origins and purposes remain a mystery, which infects your view of algebra and even of mathematics.

School mathematics is quintessentially about expressing generality, to the extent that a lesson without the opportunity for learners to experience and express a generality is not a mathematics lesson. Every method (as George Eliot pointed out), every technique, every concept, every theorem expresses or draws attention to a generality. If learners are not struggling to express these for themselves, they will be much less likely to be able to reconstruct them at some time in the future when they need them.

All of the tactics or devices being displayed in the case studies in this chapter, such as dimensions of possible variation, invariance in the midst of change, and asking what is the same and what is different about two or more objects, are intended to prompt learners to appreciate and to express generalizations that encompass the examples provided, and to remind teachers to engage learners in these metalevel reflections. Expressing generality is as much a matter of directing attention as anything else, although it can also be seen as a process of enculturation into a mathematical practice that includes quintessentially mathematical discourse. But there is a great deal more to the matter than gargling standard phrases in comprehensible sequences.

The temptation of authors and teachers is to lay out examples, perhaps as tasks, and to expect learners to build on the experience of a succession of tasks in order to become aware of or to experience that succession. But as Kant effectively points out in his *Critique of Pure Reason,* a succession of experiences does not automatically add up to an experience of that succession.

There is documented evidence of this in many different contexts. In a professional development session, I offered the following sequence of statements, with lots of pausing so that participants would see that they were supposed to check each equation in turn, and to attend to the patterns between equations:

$1 + 2 = 3 \qquad 4 + 5 + 6 = 7 + 8 \qquad 9 + 10 + 11 + 12 = 13 + 14 + 15$

"We" (meaning I wrote what some of the participants said) had written down two more expressions. I asked for an expression of generality. One person suggested $n + (n + 1) = \ldots$ but was then a bit stuck.

It transpired that his attention was entirely on the first equation. The other expressions were not seen as part of "the experience to be generalised." Rowland, Martyn, Barber, and Heal (2000) met the same thing with preservice primary teachers. When asked to check that

$3 + 4 + 5 = 3 \times 4, \quad 8 + 9 + 10 = 3 \times 9 \quad \text{and} \quad 29 + 30 + 31 = 3 \times 30$

and then to write down a statement in words generalizing these three examples, many wrote nothing or nothing that could be construed, and some wrote a false generalization such as three consecutive numbers added together equals the product of the first two, achieved by attending solely to the first example.

It is quite likely that unfamiliarity with being asked to express a generality produced a tunnel vision effect, so that attention became focused on but one instance. The fact that the two threes play different roles could easily be overlooked, which would make the conjectured generality at least understandable. In the case of the previous sequence, progress was made by asking people to chant the equations out loud but with emphasis first on the first number, then in a second pass, on the last. As Marton noted (1997, 2000), variation has to be experienced in quick succession to be experienced as variation. Asking people what is changing and what is staying the same between several objects intended to be seen as exemplary, or asking learners to say what is the same and what is different about two or more objects, are powerful tactics for directing attention appropriately so that learners attend to exemplariness of several objects rather than just one.

Extending Definitions

It has often been said that you do not really understand something until you have extended it, generalized it, and placed it in a wider or broader context. Extending (and contracting) meaning is a ubiquitous theme in mathematics (Mason & Johnston-Wilder, 2004) that accounts in part for the power and effectiveness of mathematical thinking. The following are two examples of the theme at work.

Remainders. Case study one introduces the notion of remainders and the link with a more natural description of numbers with a given remainder in terms of one more than.... Finding remainders when dividing by 5, 10, or even 7 is relatively familiar. Finding remainders when dividing by 2.4 is not so common. What could it mean? To extend the definition appropriately requires paying careful attention to the structural use of the term "remainder" when used in its familiar form. You subtract off multiples of some number m until what is left over is less than m but greater than or equal to zero. This makes perfect sense when m takes any positive value, and it applies to any real number other than zero, not just a positive whole number, as long as you are permitted to add as well as to subtract! An alternative way to deal with negatives is to require the remainder to lie between 0 and the divisor. Note that if this is unfamiliar territory, then you will want to construct some examples for yourself, and this puts you in excellent company:

> A good stock of examples, as large as possible, is indispensable for a thorough understanding of any concept, and when I want to learn something new, *I make it my first job to build one....* Counter-examples are examples too, of course, but they have a bad reputation: they accentuate the negative, they deny not affirm ... the difference ... is more a matter of emotion. (Halmos, 1983, p. 63)

It is worth contrasting this with the principle used by Hilbert (see earlier quote).

A common use of this extended idea for remainders and divisibility is to think in terms of remainders upon dividing real numbers by 1, that is, the decimal part of a real number. Thinking of wrapping the numberline around a circle of unit perimeter displays numbers lined up that have the same decimal part.

Many teachers find the notion of the remainder on dividing a negative number by some number perplexing, apparently because they want to enforce a symmetry so that remainders of negatives are negatives of remainders, rather than extending the requirement that sufficient multiples of m be subtracted (or added) until the result is a number between 0 and m. The fact that it might require negative multiples of m is sometimes puzzling at first. Similar challenges to intuition arise when m is itself taken to be negative.

The point is that very often, first encounters with a definition involve a very restricted domain of possible variation, and also a restricted range of permissible change even within those dimensions that are perceived as variable. It is only by contemplating different dimensions and ranges that the full significance and scope of a definition emerges. One of the features of examples provided by a teacher or textbook author is that through the choice of examples they provide, they may intentionally or unintentionally restrict learners' sense of what can vary, contributing to the formation of what Fischbein (1993) called a *figural concept*—a sense of a concept based on a stylized or dominant image that fails to afford access to relevant dimensions of possible variation and associated ranges of permissible change.

This applies equally to concepts and to techniques through the collections of exercises provided (Watson & Mason, 2005).

Primality. By varying a possible dimension of variation while looking for sets of numbers closed under multiplication in case study two, the notion of prime number was extended to other number systems. Again, it is easier to appreciate the significance of something when it is placed in other or more general contexts. Multibase arithmetic was proposed for this purpose, but was transposed didactically into teaching learners to get the answers in different bases, which Tom Lehrer made great fun of (Lehrer, 1965). Appreciation does not require facility or fluency, or even competence. What is valuable is exposure to dimensions of possible variation.

Learners encountering numbers are faced with a series of extensions of their sense of what constitutes number, from whole or counting numbers to zero, negatives, fractions, reals, complex numbers, and beyond. Polynomials can also be thought of as extensions of the idea of integers, and rational polynomials as the corresponding version of fractions. Even matrices can be thought of as extensions of the idea of number. At each stage, the learner is expected to let go of previous assumptions and to focus attention on a property (for primes, being irreducible through factoring), which is then used as the basis for further reasoning. Examples that served previously (as mentioned earlier, 10 and 25 are composite in the whole numbers but prime among numbers of the form one more than a multiple of 3) turn out not to serve in a varied system.

WHAT MAKES AN EXAMPLE EXEMPLARY?

What makes an example exemplary is seeing it as a particular case or instance of a more general class of objects; being aware of what can be varied and still it belongs to the class, and within what range of values it can be varied. The invariance is the type, concept, or technique. Thus, exemplariness resides not in the example, but in how the example is perceived. It is the person who sees an object as exemplary, by looking through the particular to the general. This is done by stressing some features (and consequently ignoring others) that are treated as dimensions of possible variation. If learning is based around discerning variation, then it is vital that learners be exposed to systematic (but not excessively tedious) variation, either simultaneously or in quick succession, so that the variation is actually experienced as variation. Examples that do not confuse different parameters by displaying them with the same values are more helpful in exposing variation than examples in which the same number plays several roles, as has been remarked on in the case studies.

Using metaquestioning (Pimm, 1994) can play an important role in attracting learner attention out of the particular and onto the general. For example, using tactics such as asking "what is the same and what is different about ...?" and using language akin to "dimensions of possible variation," learners can be exposed to practices that promote mathematical thinking and understanding through the use of their own natural powers.

The claim being made in this chapter is that learners who are aware of the general class of problems about which they can be asked on an examination are in a far better position to recognize any particular problem as belonging to a type, and therefore to be able to recall appropriate techniques and methods than learners who experience each problem as novel and unfamiliar. Furthermore, working with learners on appreciating the class of problems that are exemplified by the particular problems they themselves solve does far more than train them for the examination: It builds the foundations of appreciating the significance and use of a technique, and the scope and meaning of a concept. The case studies used to illustrate and hence exemplify pedagogical–didactical tactics and themes, served as examples only if you as reader became aware of dimensions of possible variation, which turn those case studies into generic examples.

The exemplification paradox that in order for examples to be seen as being exemplary, learners need to appreciate the generality that is being particularized, but the way to become aware of a generality is through examples, is not resolved by specific teacher or learner actions in a particular sequence, because appreciation of what an object exemplifies coemerges with a sense of what it is that is being exemplified. Even Hilbert's way of working quoted earlier, of finding a single generic example to probe in order to see through to the general, has the appreciation of genericity, of examplehood, of growth with increasing familiarity with the particular.

Tactics for Exposing Examplehood

The case studies have attempted to exemplify a number of tactics that are effective in promoting appreciation of generality inherent in concepts, properties, and techniques. These include:

- *Another and another:* Asking learners to construct a sequence of three or more examples meeting specified constraints in order to prompt them to become adventurous and creative, and so to explore the boundaries of their awareness of dimensions of possible variation. This was highlighted particularly in the first two tasks in case study one and in "multiplicatively closed sets of numbers" in case study two. It can be used in many situations when you want learners to become aware of and to extend their current example space.
- *Dimensions of possible variation:* Explicitly describing each aspect of some example that could be varied and still the example would remain an example, and explicitly stating the range of permissible change of

that aspect. This was highlighted in "one more than ..." in case study one, "multiplicatively closed sets of numbers" and "Euler's Sets" in case study two, and in "more factoring" in case study four, but it applies very generally not only to generalizing examples and tasks, but to appreciating concepts as well.

- *Reversing:* Converting a task that involves using a technique to get an answer into an exploration of all those similar tasks that give the same answer. A variant is to characterize all the possible answers to the class of tasks of which the particular are exemplary (see also characterizing). This appeared explicitly in "GCD & LCM reversal" in case study three, but applies to almost any task.
- *Characterizing:* Finding a way to capture or describe all possible objects that have a specified property (see also reversing). This was used explicitly in case study two, but is implicit in case study four and indeed in almost any task.
- *With and Across the Grain:* Applied to a sequence or grid of mathematical objects, including a set of tasks or exercises, in which attention is paid to detecting and following patterns associated, for example, with familiar number sequences (going with the grain), and pausing to pay attention to the structure revealed in order to make sense of the whole, not only through expressing generality, but also by interpreting the significance of that generality (going across the grain). All the tasks in case study four were intended to draw attention to going with and across the grain. It can be used whenever you want to prompt a generalization through systematically varying one or two dimensions of possible variation, as a demonstration of ways of working that learners can use for themselves when specializing.
- *Reveal and Obscure:* Construct an example (or examples) that reveals some property the example is supposed to have as explicitly as possible; construct an example that obscures that property. This was illustrated specifically in the "reveal and obscure" task in case study one, but can be used whenever you want learners to become aware of and to extend their current example space.
- *Particular to General:* Offer some particulars in order to suggest generalization (see also with and across the grain, and dimensions of possible variation). This format was used with many of the tasks, including in case study two "extending the meaning of primes" and "multiplicatively closed sets of numbers," and in case study three in "common divisors and multiples," "odd number of factors," and "relations between GCD and LCM."
- *General to Particular:* Ask a general question in order to stimulate particularizing. This was the format used in "beyond odd numbers" and in "Euler's sets" in case study two, and in "porters" in case study three,

but it applies to many tasks that start life in the particular-to-general format. Most tasks are presented in the particular-to-general format, giving learners the impression that this is how mathematics happens. Balancing presentation in the two forms can reorient learners to become aware of both movements.

Probing Awareness

Traditional assessment tests learners' reactions to probes that exemplify a class of possible tasks. In order to complete the tasks, learners need facility and fluency with taught techniques. Asking learners to construct examples of objects that meet specified constraints goes beyond testing behavior because it reveals the scope of their awareness in terms of dimensions of possible variation of which they are aware, and corresponding ranges of permissible change.

In the context of assessing learners' awareness of case study significance, Mason (1988) proposed a framework that is also useful in assessing the scope of learner appreciation and awareness of mathematical concepts and techniques (Mason, 1989). It is based on a neo-pythagorean perspective developed by John Bennett (1956–1966). The framework proposes asking learners for examples that illustrate a concept, technique, principle, and so forth, and asking them to identify concepts, techniques, principles, and so forth, which are illustrated by particular examples given by the examiner.

FINAL COMMENTS

Terms such as *investigative teaching* (and its variants such as *discovery learning*) provoke extreme reactions in many audiences, whereas *lecturing* and *starting from the abstract* provoke similar reactions in different audiences. Neither reactions are helpful because they are based on emotive associations with general labels, rather than on precise details of pedagogic strategies. When teaching that is even marginally effective is examined closely, aspects of both pedagogic stances, of starting from the particular and the concrete and starting from the general and the complex will be found to make important contributions. Strict adherence to one format is likely to foster the pedagogic theory that this is always how things are done, whereas variation in approach is more likely to broaden learners' views of what mathematics is about, what questions it addresses, and what methods it employs (Watson & Mason, 1998).

When learners search for and find solutions to tasks on the internet, they miss out on the pedagogic point of those tasks; to encounter and experience mathematical thinking. Concentration on answers as such allows pedagogic purposes and effects to drain away. Numbers provide a rich domain of study precisely because there are so many so many relationships to discover, and so many questions that learners can ask themselves as they become familiar with processes of mathematical thinking. While experiencing a transition from arithmetic computations, through particular numbers seen as exemplars of properties of classes of numbers, to a deep and structural sense of numbers comprising number theory, learners can encounter a paradigmatic exposure to mathematical thinking. Above all, the most important theory we want learners to construct is that they do actually possess the requisite powers to do mathematics and to think mathematically. Then they can make an informed choice as to whether to develop and make use of those powers within mathematics in the future.

REFERENCES

Banwell, C., Tahta, D. & Saunders, K. (1986). *Starting points for teaching mathematics in middle and secondary schools*. Oxford, England: Oxford University Press. (Original work published 1972)

Bennett, J. (1956–1966). *The dramatic universe* (Vols. 1–4). London: Hodder & Stoughton.

Bills, L. (1996). *Shifting sands: Students' understanding of the roles of variables in 'A' level mathematics.* Unpublished doctoral dissertation. Milton Keynes, England: Open University.

Brown, S., & Walter, M. (1983). *The art of problem posing*. Philadelphia: Franklin Press.

Chevallard, Y. (1985). *La transposition didactique* [The didactic transposition]. Grenoble, France: La Pensée Sauvage.

Courant, R. (1981). Reminiscences from Hilbert's Göttingen. *Math Intelligencer, 3*(4), 154–164.

Davydov, V. (1990). Types of generalization in instruction (J. Teller, Trans.). *Soviet studies in mathematics education* (Vol. 2). Reston: NCTM.

Fischbein, E. (1993). Figural concepts. *Educational Studies in Mathematics, 24*(2), 139–162.

Flegg, G. (1985). *Nicolas Chuquet, renaissance mathematician*. Dordrecht, Netherlands: Reidel.

Gattegno, C. (1970). *What we owe children: The subordination of teaching to learning*. London: Routledge & Kegan Paul.

Goldenberg, P. (1996). "Habits of mind" as an organizer for the curriculum. *Journal of Education, 178*(1), 13–34.

Halmos P. R. (1983). In D. Sarasen & L. Gillman (Eds.), *Selecta: Expository writing* (pp. 157–186). New York: Springer-Verlag.

Kaczkowski, S. (2001). Factoring quadratics. *College Mathematics Journal, 32*(3), 203–204.

Lehrer, T. (1965). Retrieved 2005 from http://curvebank.calstatela.edu/newmath/newmath.htm

Marton, F., & Booth, S. (1997). *Learning and awareness*. Mahwah, NJ: Lawrence Erlbaum Associates.

Marton, F., & Trigwell, K. (2000). Varatio est mater studiorum [Variation is the mother of learning]. *Higher Education Research & Development, 19*(3), 381–395.

Marton, F., & Tsui, A. (2004). *Classroom discourse and the space of learning*. Mahwah, NJ: Lawrence Erlbaum Associates.

Mason, J. (1988). Fragments: The implications for teachers, learners and media users/researchers of personal construal and fragmentary recollection of aural and visual messages. *Instructional Science, 17*, 195–218.

Mason, J. (1989). Teaching (pupils to make sense) and assessing (the sense they make). In P. Ernest (Ed.), *Mathematics teaching: The state of the art* (pp. 153–166). London: Falmer.

Mason, J. (1996). Expressing generality and roots of algebra. In N. Bednaiz, C. Kieran, & L. Lee (Eds.), *Approaches to algebra: Perspectives for research and teaching* (pp. 65–86). Dordrecht, Netherlands: Kluwer.

Mason, J. (1999). Incitación al estudiante para que use su capacidad natural de expresar generalidad: Las secuencias de Tunja [Provoking students to use their natural power to express generality: Tunja sequences]. *Revista EMA, 4*(3), 232–246.

Mason, J. (2001). Tunja sequences as examples of employing students' powers to generalize. *Mathematics Teacher, 94*(3), 164–169.

Mason, J. (2002). *Researching your own practice: The discipline of noticing*. London: Routledge-Falmer.

Mason, J. (2003a). Structure of attention in the learning of mathematics. In J. Novotná (Ed.), *Proceedings of the International Symposium on Elementary Mathematics Teaching* (pp. 9–16). Prague, Czech Republic: Charles University.

Mason, J. (2003b). On the structure of attention in the learning of mathematics. *Australian Association of Mathematics Teachers, 59*(4), 17–25.

Mason, J. (2006). *Advanced Factoring*. Retrievd from http://mcs.open.ac.uk/jhm3

Mason, J., Graham, A., & Johnston-Wilder, S. (2005). *Developing algebraic thinking*. London: Sage.

Mason, J., & Johnston-Wilder, S. (2004). *Fundamental constructs in mathematics education*. London: Routledge-Falmer.

Mason, J., & Pimm, D. (1984). Generic examples: Seeing the general in the particular. *Educational Studies in Mathematics, 15*(3), 277–290.

Mason, J., & Watson, A. (2001). Stimulating students to construct boundary examples. *Questiones Mathematicae, 24*(Suppl. 1), 123–132.

Minor, L. (1988). Factoring twins as a teaching tool. In A. Coxford & A. Shulte (Eds.), *The ideas of algebra K–12: 1988 Yearbook* (pp. 192–198). Reston, VA: National Council on Teaching Mathematics.

Pimm, D. (1994). Mathematics classroom language: Form, function and force. In R. Biehler, R. W. Scholz, R. Strasser, & B. Winkelman (Eds.), *Didactics of mathematics as a scientific discipline* (pp. 159–169). Dordrecht, Netherlands: Kluwer.

Polya, G. (1962). *Mathematical discovery: On understanding, learning, and teaching problem solving* [Film]. New York: Wiley.

Polya, G. (1965). Let us teach guessing (film). Washington: Mathematical Association of America.

Rowland, T. (2002). Generic proofs in number theory. In S. R. Campbell & R. Zazkis (Eds.), *Learning and teaching number theory: Research in cognition and instruction* (pp. 157–184). Westport, CT: Ablex.

Rowland, T., Martyn, S., Barber, P., & Heal, C. (2000). Primary teacher trainees' mathematics subject knowledge and classroom performance. In T. Rowland & C. Morgan (Eds.), *Research in mathematics education. Vol 2: Papers of the British Society for Research into Learning Mathematics* (pp. 3–18). London: British Society for Research in the Learning of Mathematics.

Sangwin, C. (2003). New opportunities for encouraging higher level mathematical learning by creative use of emerging computer aided assessment. *International Journal for Mathematical Education in Science and Technology, 34*(6), 813–829.

Watson, A. (2000). Going across the grain: Mathematical generalisation in a group of low attainers. *Nordisk Matematikk Didaktikk (Nordic Studies in Mathematics Education), 8*(1), 7–22.

Watson, A., & Mason, J. (1998). *Questions and prompts for mathematical thinking*. Derby, Leicestershire, England: Association of Teachers of Mathematics.

Watson, A., & Mason, J. (2004, June). The exercise as mathematical object: Dimensions of possible variation in practice. *Proceedings of British Society for Research in the Learning of Mathematics,* Leeds, England.

Watson, A., & Mason, J. (2005). *Mathematics as a constructive activity: The role of learner-generated examples*. Mahwah, NJ: Lawrence Erlbaum Associates.

Whitehead, A. (1948). *An introduction to mathematics*. London: Oxford University Press. (Original work published 1911)

Zazkis, R. (1999). Intuitive rules in number theory: Examples of "the more of *A*, the more of *B*" rule implementation. *Educational Studies in Mathematics, 40,* 197–209.

Zazkis, R., & Gadowsky, K. (2001). Attending to transparent features of opaque representations of natural numbers. In A. Cuoco (Ed.), *The roles of representation in school mathematics* (pp. 146–165). Reston, VA: NCTM.

Chapter **4**

For the Beauty of Number Theory

Nathalie Sinclair
Michigan State University

Imagine if—instead of following the established formal mathematical structures that dictate the basic guidelines of curriculum—we based our choices on more aesthetic aspects: Which mathematics will students find appealing, wonderful, surprising and intrinsically satisfying? Why might we do this? Which topics would we choose? In this chapter, I outline some of the potential aesthetic aspects of elementary number theory and thus provide additional support to several researchers' recommendations for its increased curricular emphasis (Campbell & Zazkis, 2002).

When mathematicians want to illustrate the aesthetic appeal of mathematics, as they see it, they often turn to number theory. For many, Euclid's proof of the infinitude of primes and the Pythagorean proof of the irrationality of $\sqrt{2}$ both represent the acme of mathematical elegance (see Hardy, 1940; Wells, 1990). These proofs, and others like them, possess qualities such as cleverness, succinctness, inevitability, generativity, and perspicuity that many professional mathematicians greatly value. They also highlight some of the most powerful tools of mathematics, such as the proof by contradiction and use of parity.

The following 1989 NCTM Standards quotation also draws attention to the aesthetic potential of elementary number theory: "Number theory ... explorations have payoffs in problem solving, in understanding a developing mathematical concepts, in illustrating the beauty of mathematics, and in understanding the human aspects of the historical development of number" (p. 91). The Standards quotation emphasizes the way in which the beauty of mathematics may be found in explorations, and not only in the "museum-quality" proofs revered by professional mathematicians. Indeed, proofs are not the only mathematical entities that have aesthetic appeal; the French mathematician Le Lionnais (1948/1986) pointed out that definitions, methods, diagrams, and many other mathematical entities that are involved in the process of *doing* mathematics can also possess qualities that mathematicians find beautiful. Indeed, for the professional mathematician, the aesthetic plays a fundamental role in mathematical exploration as well as in problem posing and solving (Sinclair, 2002a, 2004).

I begin with a story told by the French mathematician François Le Lionnais (1983) about his first experience, at age 7, of a mathematical discovery. Given his age, and the kinds of mathematical entities at his disposal, it should not be surprising that his discovery involved numbers, and the kinds of relationships that have delighted many mathematicians. The young Le Lionnais was sitting alone at the kitchen table, with a pencil and some paper, not tired enough to join the rest of his family for the afternoon siesta. Seated at the table with a pencil and some paper, he wrote the numbers from 1 to 9. But instead of multiplying one number by the others, as he has been learning to do in school, he multiplied each by itself, writing the result in the row beneath.

1	2	3	4	5	6	7	8	9
1	4	9	16	25	36	49	64	81

"Suddenly," he wrote, describing his memory of the experience, "a veil lifts, allowing me to perceive in this otherwise dull alignment, a beautiful structure" (p. 12; Sinclair, trans.). But to see it, he continues, one has to consent to "an amputation" (p. 12), striking out the digits in the 10s decimal place, conserving only the units. This produced:

1	2	3	4	5	6	7	8	9
1	4	9	6	5	6	9	4	1

Le Lionnais admitted that an ordinary adult might have found the resulting symmetry (with respect to the middle number) quite banal, but, having discovered it himself, he is "thunderstruck." He felt he had entered a "vast domain where a multitude of treasures has been hidden" (p. 13). Surely, he recalled thinking, he could continue mining his beautiful array simply by multiplying the digits once again (and maybe even again), each time striking out the 10s decimal place.

1	2	3	4	5	6	7	8	9
1	4	9	6	5	6	9	4	1
1	8	7	4	5	6	3	2	9

Hélas! The symmetry had been lost. Le Lionnais was stubborn, though, certain that "chaos could not have taken over the society of numbers, which had thus far been so well organised" (p. 13). And all of a sudden, he saw it: The digits occupying the symmetric positions are complements of each other in 10: 1 + 9 = 10; 8 + 2 = 10; 7 + 3 = 10; 4 + 6 = 10. Now he can move on: What happens in the fourth row? Le Lionnais worked on his rows of digits all afternoon until he was called for dinner, finding that the sixth row reproduced the second, whereas the seventh reproduced the third, and so on.

Le Lionnais recounted this story in a prelude to his quixotic book of numbers as a way of describing how he, himself, became bewitched by the queenly domain of mathematics, the theory of numbers. Remarkably, instead of extolling the beauty of Euclid's proof of the infinity of primes or the Pythagorean proof of the irrationality of $\sqrt{2}$ Le Lionnais described the beauty he saw in the process of playing with, manipulating and transforming simple whole numbers.

When Le Lionnais looked at that second row of numbers, the row of squares, he saw something; he perceived a "beautiful" pattern possibility, audaciously ignoring the 10s decimal places. When he lifted the veil and saw the symmetric structure in the row of numbers, he knew there must be more (symmetry always *means* something, doesn't it?). So onward he went to the next row, looking for the next pattern.

An ordinary 7-year-old knows about symmetry, and even about sums with results bigger then 10. But would most 7-year-olds have noticed the symmetry of the second row or the complementary balance of the third one? What about 13-year-olds? Probably not, especially 13-year-olds, who have already mastered the rule-bound game of school mathematics. In fact, the young Le Lionnais seems to have developed an unusually acute aesthetic faculty in mathematics; he appreciated the significance of symmetry, was able to ignore stuff that got in the way of a "beautiful" pattern, and anticipated further "treasures." In short, he saw numbers as interesting and potentially satisfying objects to play with.

In this chapter, I argue that elementary number theory can provide many other students, both young and old, with particularly rich possibilities for *doing* and appreciating mathematics—the latter being an ever-present goal (see, for example, the NCTM Standards) of mathematics education, but one that is seldom explicitly targeted or prioritized (despite being implicitly and deeply linked to the thoroughly researched areas of attitudes and beliefs about mathematics). I consider in particular the history of elementary number theory, its objects and methods, and its accessible wealth of patterns and intellectual delights. Several examples from research at the undergraduate level, as well as with k–12 students, are provided as supporting evidence.

WHAT MAKES "THE BASICS" SO BASIC?

A quick glimpse at mathematics curricula around the world, or a brief conversation with mathematics teachers, would probably yield similar answers to the question of what are "the basics" in mathematics. Many believe that the strict hierarchies of mathematics itself dictate some of "the basics," whereas others find them dictated by the needs of modern society. Certainly, some components of elementary number theory, such as the study of prime numbers, fit into "the basics" of both categories. But Papert (1980) asked educators to consider the question of what's basic for whom, implying that the hierarchies of mathematics are not sufficient to answer "the basics" question. He hinted at another, more psychological and less mathematical, way of answering the basics question by appealing to aesthetic criteria, criteria mathematicians use all the time to answer questions about significance and worth (Tymoczko, 1993). The question of the basics should keep in mind the kinds of mathematics children are good at, as well as the kinds of mathematics that children enjoy and find interesting. With such criteria in mind, a topic such as fractal geometry may have greater warrant for inserting itself in the school mathematics curriculum (see Goldenberg, 1989). In terms of elementary number theory, I propose several characteristics that qualify it as part of the basics, using a Papert-inspired expanded set of criteria. They are as follows: Children *have* numbers, children seek pattern, and, elementary number theory provides access to the fundamental values and insights into the power of mathematics.

The first characteristic reflects children's deep and long-standing familiarity with the building blocks of elementary number theory; the natural numbers. It is with these mathematical entities that children have come closest to achieving the kinds of experiences and connections that characterize mathematicians' relationships with entities such as derivatives, groups, and matrices. The second characteristic reflects basic human ways of learning, by anticipating, looking for, and identifying patterns. And elementary number theory is chock full of the kinds of patterns that children are able to notice, in part because these patterns involve the mathematical entities with which they are most familiar. Being able to notice patterns is fundamental to problem posing and solving, as Le Lionnais' episode illustrated so well, and these are arguably the most important activities of mathematicians. Last, elementary number theory offers the most accessible examples of the fundamental values and processes of mathematics including such things as classification, generalization, precision, and economy across the grades. Kindergarten students can grapple with the idea of infinity; second grade students can grapple with the question of whether two even numbers always sum to another even number; seventh grade students can consider methods for identifying prime numbers; 10th grade students can appreciate the precise way in which it's possible to show there are infinitely many primes.

AN AESTHETIC PERSPECTIVE ON "THE BASICS"

Before considering each of the three characteristics outline earlier, I would like to describe very briefly the importance and nature of the mathematical aesthetic, if only to provide some categories and vocabulary that will be used throughout the rest of the chapter.

Recognition of the beauty of mathematics (and claims about it being the purest form) is almost as old as the discipline itself. The Ancient Greeks, and particularly the Pythagoreans, believed in an affinity between mathematics and beauty, as described by Aristotle; "the mathematical sciences particularly exhibit order, symmetry, and limitation; and these are the greatest forms of the beautiful" (1966, XIII, 3.107b). Many eminent mathematicians have since echoed his words. For instance, Bertrand Russell (1917) wrote that mathematics possesses a "supreme beauty ... capable of a stern perfection such as only the greatest art can show" (p. 57). G. H. Hardy's (1940) words show slightly more restraint in pointing out that not all mathematics has rights to aesthetic claims: "The mathematician's patterns, like the painter's or the poet's must be beautiful; the ideas, like the colours or the words must fit together in a harmonious way. Beauty is the first test: [T]here is no permanent place in this world for ugly mathematics" (p. 85).

These descriptions will almost certainly resonate with mathematicians, but they may do precious little to enlighten the nonmathematician: What do aesthetics have to do with the way that mathematicians actually work and with the way that mathematics evolves? My attempts to answer these questions have helped me identify three groups of aesthetic responses, which play three distinct roles in mathematical inquiry. The most recognized and public of the three roles of the aesthetic is the *evaluative;* it concerns the aesthetic nature of mathematical entities and is involved in judgments about the "beauty," "elegance," and significance of entities such as proofs and theorems. The *generative* role of the aesthetic is a guiding one, and involves nonpropositional modes of reasoning used in the process of inquiry. I use the term generative because it is described as being responsible for generating new ideas and insights that could not be derived by logical steps alone (for example, Poincaré (1908/1956)). Lastly, the *motivational* role refers to the aesthetic responses that attract mathematicians to certain problems, and even to certain fields of mathematics. I argue in the next few sections that elementary number theory can provide a mathematical context in which learners can appreciate and activate these three roles of the mathematical aesthetic. This aesthetic dimension of elementary number theory should been seen as augmenting the mathematical and philosophical ones Campbell proposes in his chapter (this volume).

WHY ELEMENTARY NUMBER THEORY COUNTS AS "BASIC"

In this section, I draw on several research examples in order to both illustrate and support the characteristics of elementary number theory described previously.

Children Have Number

Nonnegative integers—the primary objects of elementary number theory—are pervasive in children's lives, as are the operations of multiplication and division. These are objects and operations that children see, know, and use "... students from middle school through college feel comfortable dealing with whole numbers" (Selden & Selden, 2002, p. 214). This level of comfort and acquaintance can help in conjecturing and problem solving, as it did with the young Le Lionnais: Students working with number theory situations or activities have the cognitive liberty of concentrating on exploring, guessing, and "playing with" without being distracted by having to simultaneously extend their conceptions of mathematical operations and objects. Silver and Metzger (1989) have shown how this cognitive liberty is crucial to enabling aesthetic, mathematical thinking—even in "expert" mathematical activity.

A higher level of comfort and acquaintance can also stimulate student motivation, particularly with respect to the potential for surprise and paradox: The more you know about something, the more you will expect certain regularities based on your experiences, and the more likely you will be surprised by contradictions or interruptions. Zazkis (1999) illustrated a simple example of potential surprise with respect to the number for factors of an integer F(n): Students expect that the greater A is, the greater F(A) should be (or if $A_1 > A_2$ then $F(A_1) > F(A_2)$), using the reasoning that works well in everyday contexts that the more of A, the more of B. Students are surprised by mathematics when their expectations, even those based on their intuitions, are contradicted. For instance, students can be surprised that a small change can make a big difference, that a random collection of objects can share a common property, or that analogies can prove to be nonanalogous. Moreover, as Movshovits-Hadar (1988) argued, surprise can provide exciting experiences for students in the mathematics classroom. The feeling of surprise stimulates students' curiosity, which can, in small steps, lead toward intelligibility; it makes the students struggle with their expectations—and with the limitations of their knowledge—and thus their intuitive, informal, and formal understandings; and lastly, it provokes students to develop an appreciation of the significance of the concept in question.

Surprise and familiarity played important roles in the problem-posing and problem-solving processes of middle school students working with the Colour Calculator (CC), an internet-based calculator that displays the decimal portion of real numbers in a color-coded grid of changeable width. The research was conducted with 15 grade-eight students (eight boys and seven girls) of mixed ability (as rated by their regular classroom teacher). The students came from lower to middle class, small town backgrounds in North America. Before describing the research insights (for a more complete report, see Sinclair, 2001), let me first describe the microworld.

The calculator operates at a maximum precision of 225 decimals digits, and thus, of particular interest in the CC are the pattern-rich real numbers because they can be seen and understood as patterns of color. This graphical representation of number calls attention to and facilitates the perception of important subsets of real numbers—terminating, periodic, eventually periodic and nonperiodic—and some of their properties. The calculator operations (addition, subtraction, multiplication, division, and square root) as well as the changeable width of the table enable exploratory action on the color patterns themselves. In Fig. 4.1, the operation 1/7 has been typed into the calculator with number of decimals displayed set at 100, and table width set at 10. The CC generates the associated colored table.

Using the button that controls the width of the table of colors, the student can select different table dimensions that result in different color patterns, some that highlight interesting aspects of the fraction's period. Figure 4.2 shows 1/7 displayed using a table width of 18 and 17, respectively.

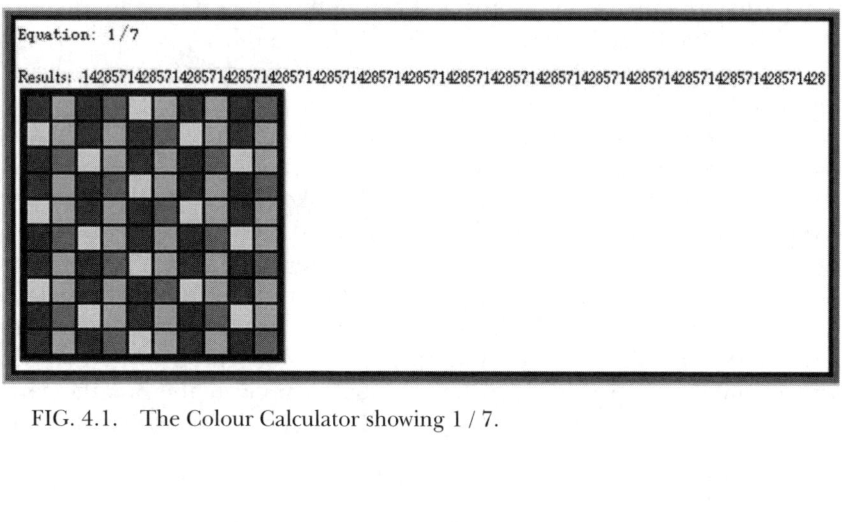

FIG. 4.1. The Colour Calculator showing 1 / 7.

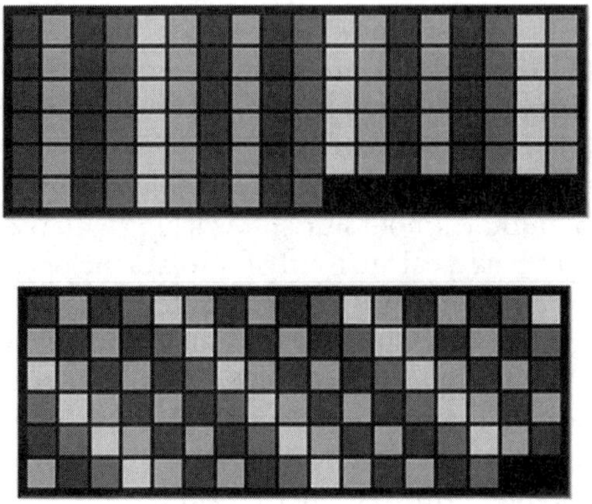

FIG. 4.2. Different representations of 1 / 7 in the CC.

Given this basic environment engaging the visual capacities of human perception, what kinds of mathematical observations emerge? The types of questions that seem compelling to ask in this environment relate mostly to patterns in the decimal expansion of real numbers: What different types of patterns exist? How does the pattern depend on the number? These questions can lead to sophisticated number theory questions, especially with prime numbers, that many mathematicians find compelling.

However, they are not all necessarily the type of questions or properties compelling or suitable to middle school students. But by beginning with this aesthetically rich situation and asking what is interesting or significant to know about fractions and their decimal representations, a radical departure from conventional curricular objectives might ensue. For instance, rather than emphasizing the importance of understanding fractions as parts of wholes or of knowing how to convert fractions to decimals, the CC emphasizes the understanding of fractions as patterned objects (as representations) and the relationships between numerators and denominators that determine these patterns.

In order to help evoke aesthetic perceptions, the environment is structured to offer only minimal instructions and a few specific questions. In particular, in order to draw students into the pattern-rich rational numbers, they are asked to begin by trying a few fractions and observing the associated table of colors. The first fraction they are asked to try is 1/7, which gives a repeating, nonterminating result. They are also encouraged to change the width of the table, and then to continue experimenting with other fractions. These minimal explanations present the students with the potential for encountering initial complexity—they have to make sense of the colors associated with 1/7, to connect the fraction to the legend provided on the screen; and that complexity has the potential to motivate their mathematical activity. In fact, Brown (1993) pointed out that the structure of traditional classroom tasks actually gives students a false sense of aesthetic unity. By presenting students with tasks that are always as clear, coherent, and efficient as possible, they are deprived of the possibility of feelings of confusion, doubt, ambivalence, and dissonance, the very feelings that produce the tension and frustration necessary for mathematical inquiry.[1] Without confusion or disharmony, students do not have the opportunity to bring their cognitive, aesthetic, and affective structures to "good form." After all, it is the reduction of surprise and complexity to simplicity and predictability that evokes, in learners, first reflection, then pleasure. Thus, the apparent insufficiency of textual instructions sets up from the outset, trajectories of sense making ranging from complexity/confusion to simplicity/pleasure.

After reading the instructions, the students started working while I asked some questions to try to get to some of the reasons behind their actions. At an obvious and almost trivial level, every student expressed that they had never seen fractions or decimals like this—many of them realized for the first time that a fraction and its corresponding decimal representation are the *same*[2]—one student noting "you never see them together like this." Every student also expressed how different this type of mathematical activity was from their regular classroom work, one explaining "you actually have to *do* things" whereas another observed (in words close to Mason's heart) that "you have to *notice* things."

Of the 15 students I interviewed, 13 of them showed obvious physical signs of surprise, which they expressed through one or more of the following actions; widening their eyes, sitting upright or moving forward, making a sound such as "ooh," or saying some form of "wow." Most were surprised that numbers could be displayed in this colorful way, but some were more surprised by the appeal of the colored pattern, much as they might show surprise encountering a colorfully tiled ceiling. Whatever the source, their initial surprise caused them to want to know where the colors came from and whether other fractions would produce similar results, as indicated by their subsequent actions (asking me, trying another fraction, studying the legend).

One student, whom I will call Nadia, showed no physical surprise at all, and answered "I don't see anything" when I asked her what she saw in the table of colors. Nadia was either completely insensitive to the patterns in the table or, due to her timidity and lack of confidence, she may have been un-

---

[1]Stigler and Hiebert (1999) made a similar point in comparing the United States with Japan with respect to TIMMS results in light of curriculum and textbook approaches in the two countries.

[2]It is interesting to recall that a regular calculator replaces its input with its output so that a student calculating 1 / 7 on the calculator never actually sees both the fraction 1 / 7 and its decimal expansion simultaneously. Although I am sure that the students think that a fraction and its decimal representations are equal, they seemed struck here for the first time by an ontological equivalence.

der too great of an affective barrier to even attempt to engage. (She may also have been at least partially color-blind). The other student who showed no physical reaction was Cameron, who remarked flatly, "there are lots of colors and patterns there." I remember feeling a bit disappointed at Cameron's remark. Cameron is a good reminder that surprise is subjective; note that most students are utterly unsurprised that the three medians in a triangle always meet at a single point, or that the exterior angles of an convex polygon always sum to 360°.

The aesthetic response of surprise may have drawn most students in, but I was more interested in finding out what their sense of surprise prompted them to perceive, and what kind of actions those perceptions entailed. This was easiest to observe with the more articulate students who provided a running commentary of their thought processes, like Sean:

> Okay. Ah. It looks like an abstract painting. Not exactly like a math problem. I'm trying to figure out how it calculates that. Uh. Well, it says that the results are 0.142857 and it repeats. So this is a repeating pattern. I can see it because the red sticks out and the purple, and ooh the green. They kind of go in a diagonal that shows a standard repeating pattern but I'm trying to figure out how things are working. So the number corresponds to the color ...

Here Sean is not only noticing particular colors—he seems especially to like green—but a repeating pattern that draws him in, making him want to understand where the pattern comes from and "how things are working." He first realizes that each color corresponds to a number, and then, having noticed the repeating pattern, wonders whether there will always be a pattern. He tries several other fractions and eventually decides to investigate whether he can find a number that has "no pattern." Sean is looking for the pathological example; he wants to find the fraction that breaks the repeating pattern rule.

On the other hand, Ann's first observation is that "every seventh box is a purple." She also sees the repeating pattern but because she focuses on the last color of the repeating sequence, purple, she forms a more quantitative interpretation of the situation, noting that 1/7 has six repeating digits (its period is six). She "verifies" this by changing the width of the table from 10 to 6. She then begins to investigate the periods of other fractions. Anne wants to see whether there is a relationship between fractions and the periods of their decimal expansions.

After seeing the result for 1/7, Kirk immediately tries 1/3, and then ½. He focuses on the differences between the table of colors that each fraction produces, noting that some terminate whereas others do not. He begins to categorize various fractions as terminating and nonterminating and then tries to figure out how he can predict to which category a fraction belongs based on its denominator. As was the case with Zoe, Kirk's urge is to categorize, upon perceiving that fractions can be divided into two distinct groups.

Julie takes a slightly different approach; she begins by changing the width of the table of colors several times. When the width is set to 7, she remarks "it's like a staircase." When the width is set to 3, "it's doubled up." Julie is interested in describing all the types of possible patterns 1/7 could produce. She seems particularly interested in the diagonal pattern and experiments with other fractions, trying to figure out how the grid width should be set in each case. Julie wants to see fractions as diagonals. (Interestingly, most adults prefer to find table widths that create stripes; perhaps stripes are just a bit too regular for Julie, diagonals providing some not-too-complex but welcome deviation for the teenager.)

Three other students embarked on explorations similar to the three described earlier. Each of these seven students in total perceived the CC's output differently, some focusing on differences, others on similarities, and others on specific patterns. Their individual perceptions, rooted perhaps in the initial surprise or appeal, gave rise to different questions, problems, and investigations. But each inquiry was motivated by an aesthetic urge: Sean was most "romantic" in his desire to find the rebellious case, the bizarre and nonconforming fraction. Ann and Kirk were both more "classical" in their orientation, wanting to find some structure or order. Julie wanted to see whether she could impose a certain, personal form on each fraction.

Table 4.1 presents a list of student investigations, in their words, along with a corresponding version reformulated as a question in more conventional mathematical discourse. Each student question reveals a distinct perception that ultimately shaped the actions taken by the student, and thus their paths of inquiry.

From the questions listed in the table, it is quite obvious that the students investigated and learned quite a few things about fractions and their decimal representations, as well as elementary number theory. For example, in order to get a display of diagonals or stripes, the students had to consider the concepts of factors or divisibility. Students were engaged in problem posing, they experimented and used problem-solving strategies, and they expressed their findings to me: They were doing mathematics. However, I want to draw attention to the fact that each problem in the table grew out of an aesthetic engagement with the CC, first through surprise and appeal, which then led them to form various perceptions of the color calculator's output. Based on these perceptions, students developed interpretations and conjectures on which they could act through experimentation with the various functions of the CC. Thus, the appeal and surprise motivated *and* shaped the mathematical activities of the students.

TABLE 4.1
List of Student Investigations

Student question	*Reformulation*
How can I make the diagonals go in the opposite direction?	How are diagonal patterns related to the period?
Maybe the period is always one less than the denominator.	How is the period of the fraction related to its denominator?
I think that all the fractions with odd numbers on the bottom will repeat.	What values of the denominator yield non-terminating decimals?
The decimals stop when there is a 2, 5, or 10 in the denominator. I wonder what happens with 8.	What values yield terminating decimals?
So with 12 on the bottom, there's an extra number before the repeating.	When is a decimal eventually periodic?
I want to take away those colors that don't fit with the rest of the pattern.	How do you move the decimal point to the right?
Aren't there some numbers that have a totally random pattern?	What kinds of numbers neither terminate nor repeat?
I want to get the table all red.	What fractions have a repeating expansion of one digit?
I wonder what happens if I try 3 / 7 now.	How are fractions that have the same denominator related?
So what do I get if I add 1 / 2 to this 1 / 3?	What is the effect when you add two fractions with different periods?
I think that since 1 / 9 has a period of 1, like 1 / 3 then 1 / 49 should have a period of 6 .	What is the effect when you square a fraction that has a certain period?

Related to the "children have number" theme I have been developing in this section is the notion that humans have had number for a long, long time. Mathematicians are often aesthetically motivated by a feeling that mathematics is somehow transcendental—providing them with a sense of order and certainty in our otherwise chaotic world. Similarly, there is an appeal to the seemingly transcendental types of problems and concepts that have been with us—or could have been with us—since the time of the Ancient Greeks: Mathematicians continue in the tradition of Pythagoras to look for proofs of the irrationality of $\sqrt{2}$ (see, for example, Apostol's 2000 proof); many questions about prime numbers, factoring, and partitioning are as vibrant today as they were 2,000 years ago. The realization that she is posing a problem, or working on one, that Zeno or Euler could have posed might stimulate in a student an aesthetic motivation much more profound than the purported motivation factor associated with the "real-life" problems populating so many textbooks and curricula. With the CC, the students did not pose a question as famous as Zeno's, but some of their questions have been posed, and solved, by other mathematicians.

Children Seek Patterns

The types of problems that crop up in elementary number theory are, as Ferrari (2002) noted, particularly well suited to those actions of pattern discernment and exploration that are eminently available to students. Humans are good at patterning—it is how they learn about and appreciate their world (Gombrich, 1979). When students recognize that they can activate a behavior they know they are good at, they can derive the same kind of aesthetic appeal that motivates mathematicians to seek out and take on the types of problems that activate the ways of thinking with which they have had success—whether it's symbolic juggling or visual representation. Also, many of the problems and objects of elementary number theory can be patterned in concrete, accessible forms (for example, actual numbers as opposed to, say, abstract algebra objects; the patterned numbers in the Colour Calculator; figurate numbers; Pascal's triangle[3]) on which the qualitative, aesthetic ways of reasoning can more easily operate. Finally, with such pattern-rich, accessible situations, students can undertake that most crucial of mathematical activities; problem posing (Brown & Walter, 1983). It is with the act of posing problems—deciding whether they are worth pursuing and convincing others that they are—that the evaluative characteristic of aesthetics is first operative for mathematicians, and can be first stimulated in students.

I would like to introduce another computer-based environment that makes the patterns of elementary number theory accessible to students, both through visualization and experimentation. Research using *Number Worlds,* a microworld I describe shortly, with elementary preservice teachers showed very positive results in terms of inciting students to explore, helping students to overcome long-standing, misleading, and difficult-to-displace intuitions, and providing them with examples, spaces, and visual images of concepts that improved problem-solving abilities (see Sinclair, Zazkis, & Liljedahl, 2004). This research also showed how compelling number patterns could be for students, and how these patterns helped students appreciate the importance and prevalence of relationships in mathematics.

---

[3]In fact, Hawkins (2000) argued that by working with a combination of number and form—of the spatial and pictorial with the analytic means of understanding—that students' aesthetic engagement can best be stimulated and supported.

Figure 4.3 shows a snapshot of the Number Worlds applet.[4] The center grid contains a two-dimensional array of clickable cells. The numbers shown in the cells depend on the "world" that has been chosen. Although the basic objects of Number Worlds are the positive integers, or natural numbers forming the Natural World, the user can choose among other sets; the Natural World, Whole World, Even World, Odd World, and Prime World.[5] It is also possible to change the numbers shown in the cells. The user can increase or decrease the start number by one row, that is, by the value of the grid width, or simply by one cell. The Reset Grid button resets the start number to the first number of the current world.

The appearance of the grid can be affected by changing the value of the Grid width menu. By selecting values from 1 to 12, the user can change the number of columns displayed, and thus the total number of cells. There are always exactly 10 rows. Within each world, the user can highlight certain types of numbers: Squares, Evens, Odds, Primes, Factors, and Multiples. Further, the multiples that are chosen can be shifted by an integer in order to create any arithmetic sequence. Finally, the four basic arithmetic operations are available to the user. The result of the operation, in addition to the two inputs, is highlighted on the grid while a string representing the operation and the result appears above the grid.

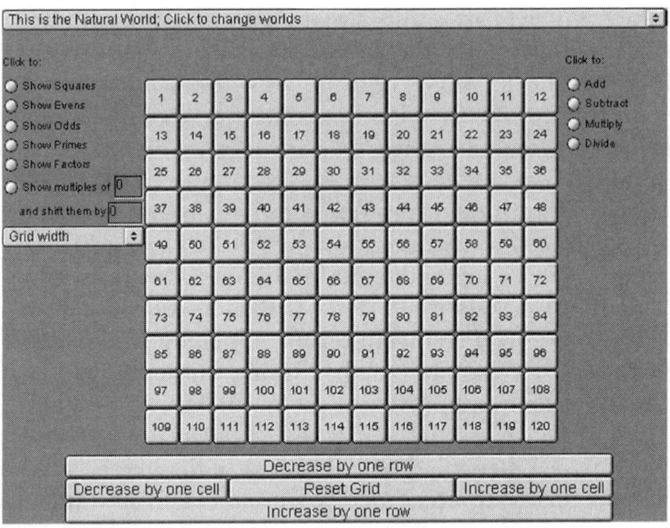

FIG. 4.3. The Number Worlds microworld.

[4]The Number Worlds microworld was written in Java and is available online as a applet at http://www.math.msu.edu/~nathsinc/java/NumberWorlds/.

[5]This approach was inspired by Brown (1978), who uses similar ways of re-examining ideas in elementary number theory by shifting the focus from natural numbers to other domains.

Although many may already be able to imagine the types of patterns that can be made and the range of mathematical ideas involved in creating and understanding these patterns, I have described several in Appendix A. As the research participants began working with Number Worlds in the lab, they seemed engaged and explorative. Many responded with surprise and even delight on creating their first patterns of stripes and diagonals (see Fig. 4.4).

Both in their written assignments and during their interviews, the students had opportunities to qualitatively describe their experiences working with the elementary number theory ideas inherent in Number Worlds. They were explicitly invited to do so in their final assignment and in their interview. Two noteworthy themes related to aesthetics emerged; the *appeal* of the patterns, and the importance of *connections*. Interestingly, these themes have a strong overlap with Burton's (1999) epistemological model that describes the way in which mathematicians come to know mathematics.

Appeal Theme. Many of the participants, both in the lab and during the interviews, commented on the appeal of the diagonal patterns produced by certain inputs to "show multiples of." Whereas stripes—as shown

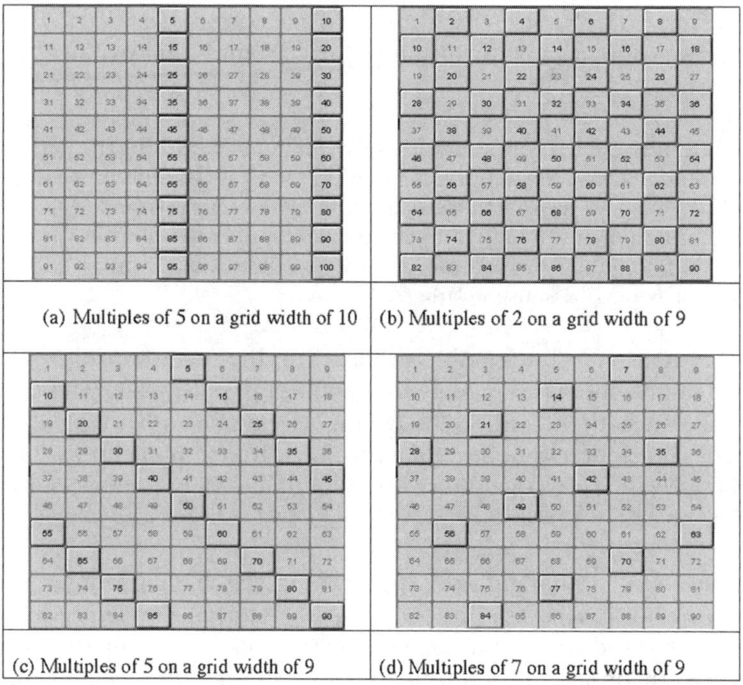

(a) Multiples of 5 on a grid width of 10

(b) Multiples of 2 on a grid width of 9

(c) Multiples of 5 on a grid width of 9

(d) Multiples of 7 on a grid width of 9

FIG. 4.4. Some possible patterns in Number Worlds.

in Fig. 4.4a, which occur whenever the show multiples of is a factor of the grid width—were deemed "boring" by many, diagonals—as shown in Figs. 4.4b and 4.4c—were evocative of patterns and relationships while possessing a pleasing degree of complexity:

> Multiples can create some visually pleasing stripes and diagonals along with some eyestraining checkerboards. There is a sense of order in the universe upon viewing some stripes on the grid. I look at those stripes and think that I can instantly recognize this regular pattern as a logical pattern.

Whereas one student described the stripes as conveying a "sense of order," another student conveyed a more emotional experience, resonant with the kinds of descriptions the young Le Lionnais offered: "The repeating, infinite pattern is calming. I can see with my eyes open the pattern on the grid right in front of me. They all seem to be linked together, no matter how far away from the next multiple they are, they are all connected to each other." Other research participants spoke of the patterns being "aesthetically pleasing," perhaps in part because of their constancy and predictability: "[A]ll the visual images ... are constant and predictable and they are aesthetically pleasing in their design or pattern."

More importantly, the general appeal of the patterns seemed to have encouraged most participants to simply "play around" and explore, as Lisa explained.

> It was fun, just playing around with different numbers, seeing the patterns you could make, like often times I wouldn't look at the numbers, I would look at rows and look at patterns, and how those corresponded to each other and then put the numbers in afterwards, so experimenting around with that, I could figure out different ways of getting the things without necessarily having to use numbers.

The participants' aesthetic response to the microworld's patterns seemed to give them access to greater expressive power, allowing them to communicate both their ideas and feelings through Number Worlds. Their aesthetic connection went beyond a frivolous attraction to patterns: It provided them with a new, qualitative way of thinking about numbers and their relationships.

Connections. Noticing and appreciating the connections in mathematics constitutes a significant part of mathematicians' pleasure in doing mathematics, and acts as a fundamental criterion for judging the worth of new ideas, theorems, and even fields. Noticing and appreciating connections is also frequently touted as an important goal of mathematics education, and one that has featured strongly in the NCTM Standards as well as most of the reform curricula. Often, educators focus on evoking the connections between mathematics and "real-life," and sometimes, they try to draw students' attention to the connections between distinct curriculum areas, such as geometry and algebra—and this to limited success. However, many of the participants described the way Number Worlds allowed them to see relationships, and to see how different concepts connected. Connections could be between different concepts as it was for Lisa, who said that Number Worlds "helped me understand things better, ... it just kind of connects different parts of multiples and math and stuff like that altogether." Connections could also simply be within the objects and operations of Number Worlds. For example, Jessica talked about the way the visual display emphasized connections between the grid's numbers: "I don't know, it was neat to see for the factors, like the patterns you can make with all the different ways, like checkerboards and all that kind of stuff, it was neat that numbers can actually do that when you put them all together, like that." Katherine emphasized the way in which the visual image helped forge connections for her:

> I find it interesting, the whole idea of like the cells highlighting and see how they relate and just like even looking at these pictures too, kind of making everything back to the visual image that it created and how they kind of related to each other and how it's like not prime or how it is prime, because of it.

Carrie explained how Number Worlds emphasized relationships for her because it got her "into a different way of thinking about it, like it's their relationship, it's numbers and their relationship to each other that I found interesting, visually." She also explained that these relationships were more clear on the Number World's grid than on a number line: "if you do a number line, every even number is, you see it in a line, but somehow in a pattern you really see just the relationship to each other."

Finally, connections could be between mathematics and the outside world, as it was for Vanessa, who talked about how her Number Worlds experience has changed the way she thinks about numbers in her everyday life.

> I will definitely take what I've learned about manipulation of numbers and seeing patterns with me as I use numbers in other contexts of my life ... I definitely learned a lot of different ways of looking at numbers, I don't just, like I'm looking at the number 64 right now, before I was just kind of taking advantage of, of it being 64, but now I kind of think how is that related to the other numbers that are in the set of numbers that I'm looking at.

Based on Devlin's (2000) characterization, her statement bears striking resemblance to the way mathematicians relate to mathematical objects; "mathematicians think about mathematical objects using the same mental faculties that the majority of people use to think about other people" (p. 262). For mathematicians, numbers are like "characters in the mathematical soap opera": but, instead of wondering about marriages and love affairs, mathematicians wonder about mathematical facts and relationships. How is this number related to that one? What kinds of number have a certain property? In a similar vein, the calculating wizard Wim Klein remarked "Numbers are friends to me." Taking 3,844 as an example, he said, "For you it's just a three and an eight and a four and a four. But I say 'Hi, 62 squared!'" If being able to do mathematics is indeed related to an ability, or a willingness, to see numbers as "friends," or as entities that can have personalities, it seems that explorations in elementary number theory, using these types of visual, experimental environments, could provide excellent "basics" for students.

A Mathematical Showpiece: Windows Onto Fundamental Values and Power

Geometry has traditionally been the topic area through which students are exposed to that ultimate mathematical entity of proof (and its related arts of conjecturing and verifying), but apart from an exposure to the various proofs of the Pythagorean theorem, maybe, they have few opportunities to respond to the various elegant, brutish, insightful, practical ways that mathematicians have of explaining and convincing. However, through elementary number theory, even students with little mathematical background can encounter many of these methods such as proofs by contradiction, induction, generic examples (see Rowland, 2002), visual proofs, and counterexamples. As Stephen Brown (1973) has shown, students are quite able to engage in the kind of aesthetic valuing that characterizes mathematicians' evaluations of proofs and solutions. Brown asked his students to generate methods of summing the integers from 1 to 100. Although Brown's students responded most strongly to the degree with which a solution to a number theory problem revealed their own path through the problem, its *genealogy*, they could also appreciate, for instance, the efficiency of a solution, the cleverness of a proof, and the perspicuity of a diagram; and, these are all criteria that mathematicians such as Hardy (1940) have proposed for judging the worth of mathematical entities.

Part of the accessibility relates to the "well-known 'folk theorem' among mathematicians that number theory questions and conjectures can be easy to state but phenomenally hard to prove" (Selden & Selden, 2002, p. 214). For mathematicians, this is a frequent source of aesthetic motivation: Questions that are easy to state can elicit in them not only the same sense of wonder that Aristotle claimed grips philosophers, but can provide them with a way of connecting with nonmathematicians and mathematicians alike—everyone can understand Goldbach's conjecture! In a related way, elementary number theory features many of the aesthetic wonders of mathematics—ideas such as infinity and scale—that Egan (1990) has shown have intrinsic appeal to students. Such ideas can thus initiate Whitehead's (1929) romantic stage of mathematics learning during which the "subject-matter has the vividness of novelty; it holds within itself unexplored connexions with possibilities half-disclosed by glimpses and half-concealed by the wealth of material" (p. 29). Whitehead argued that this stage was crucial and necessarily antecedent to the stages of precision and generality into which students are often prematurely pushed.

Deborah Ball's research shows that even grade 2 children can engage in some of the most important mathematical processes using elements of elementary number theory. In one particular example, the students invent a new kind of number, which they term "Shea numbers," after their inventor Shea; these are numbers that have *both* even and odd factors. Shea numbers evolve from classroom debates over whether 6 should be called even or odd. After Shea is finally convinced that 6 is indeed even, despite the fact that it has an odd factor of 3, the students realize that it isn't like a large class of "really odd" numbers such as 5, 9, 15, and so forth, which only have odd factors, nor is it like a large class of "really even" numbers, such as 4, 8, 16. So they invent a new kind of number, using a similar process, and motivated by comparable aims, that characterizes the most important developments in mathematics, including that of prime numbers, rational numbers, and real numbers.

It is perhaps telling that Tymockzo turned to number theory as well when trying to illustrate the way in which proofs can be analyzed and interpreted in much the same way that art critics approach works of art. He offered an interpretation of the fundamental theorem of arithmetic (which states that every natural number greater than 1 has a unique prime factorization) that is meant to be experienced "in a way that one experiences a (short) piece of music or a poem." It exists "in time with a tempo or density, speeding up at certain points and slowing down at others." It attends to the "different rhythms in it, now quickly and easily moving along, now encountering a stumbling block, tension or subsidiary tension." Also the narrative has a certain memorable quality in it. It is coherent. It hangs together. It is easier to remember because the steps have been colored differently, some highlighted, some forced to the background. Elsewhere, I have argued that such interpretations are woefully absent in school mathematics, yet potentially pivotal in providing students with models of mathematical narratives, and their associated powerful ways of thinking.

Once again, for the very reasons that "children have number," elementary number theory can provide an excellent domain not only to teach proof, as Rowland (2002) has argued, but also to introduce students to the kinds of values that govern the very form of proof, values such as precision, generalization, stability, and certainty. Because proofs are not usually the primary currency of mathematics students, solutions to problems can be used to evoke similar kinds of values. By returning to the CC, I can provide a good example of this. Here, two middle school students are asked to compare their solutions to the problem of finding a table of color that begins orange, blue. Instead of asking the students to explain their solutions to each other—a frequent request in the mathematics classroom—I ask them to consider which solution is better. Of course, this requires some explanation, but the explanation has a persuasive function; it is not merely undertaken to satisfy the teacher's request. The ensuing negotiation illustrates that students take the value question seriously, and have some strong aesthetic preferences when it comes to their mathematical evaluations, some of which are perhaps surprisingly contiguous with those of the mathematical community.

After many tries, John figures out how to create a table of colors that starts orange, blue using the fraction 72/99. Sitting behind him, Nora also hits on a solution to the problem; 72/100 (see Fig. 4.5). Seeing both of these solutions on their screens, I draw them to the students' attention. When John sees Nora's screen, he hits his forehead, and says: "oh, I should have thought of that ... I forgot but it's just going the other way isn't it?" Nora

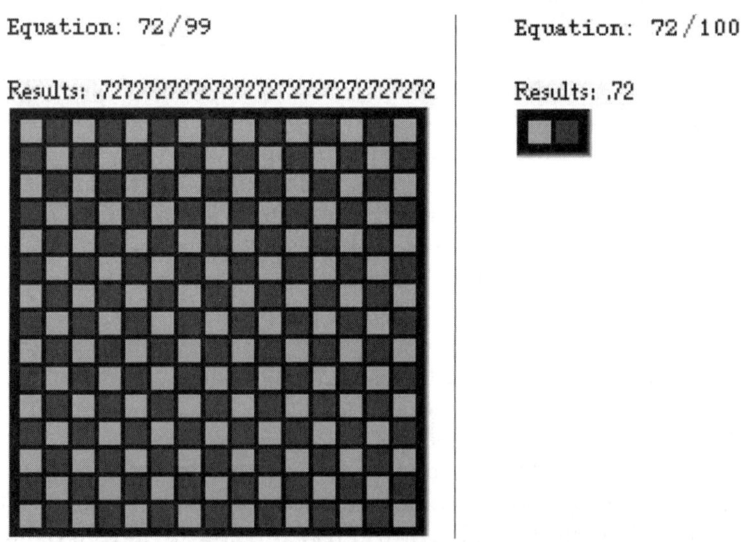

FIG. 4.5. John's solution (on left) and Nora's (on right).

raises her eyebrows when she sees John's screen. I ask them which solution is better.

John argues that his fraction actually gives him a method to get any initial sequence of colors; if he wants a decimal expansion that begins with 338, for example, he can use his "9's method" to come up with a correct solution of 338/999. Nora counters that her method is just as generative; an initial sequence of 338 can be provided through her decimal-based method with the fraction 338/1000.

Then John responds that his fraction is "nicer looking," because it produces a whole table of repeating colors, rather than a small sequence, and that it is "cooler," because it uses the properties of "my magic number." This seems to convince Nora, because she turns around and starts experimenting with John's method.

With little prompting, Nora and John negotiate the value of their respective solutions using several different criteria. The first criterion John invokes is that of *generativity;* he has realized that his solution reveals a generalizable pattern and he seems to value the idea that his particular solution can be extended to have a more general application. The appreciation of generativity, which both Nora and John show, is actually quite mathematically sophisticated. The mathematician Hardy (1940) mentioned it explicitly in his discussion of the criteria for mathematical beauty, citing as an example the Pythagorean proof that $\sqrt{2}$ is an irrational number (a proof that can easily be extended to other prime numbers).

In addition to the criterion of generativity, John also invokes a more personal criterion—his solution uses a *novel* method that he himself discovered—that might have better encapsulated his process, much like in Brown's example. When I saw his reaction to Nora's solution, a groaning recognition of his knowledge blip, I thought he would prefer hers. But John clearly treasures his magic number 9, and his solution exploits properties he has newly discovered. In contrast, Nora had found her solution 72/100 quite easily and was perhaps not as invested in the properties she had exploited as John was in his. She might also have failed to recognize the extent to which her method uses the not-so-magical, yet still quite solid, and easy-to-explain decimal place value system. Neither John nor Nora placed much emphasis here on the criterion of transparency; they both end up preferring John's method even though they might not be able to explain why it works, whereas they both understand why Nora's method works.

And finally, Nora and John both subscribe to the criterion of *visual appeal;* indeed, this seems to be the criterion that ultimately convinces Nora. Mathematicians rarely mention this criterion, perhaps because they communicate most of their work in nonvisual modes (and are somewhat suspicious of relying on visual supports to reasoning). This might be changing somewhat, given the advent of powerful computers, as illustrated in the well-respected mathematician David Mumford , Series, and White's recent colorful book (2002), which uses carefully chosen and aesthetically manipulated images to communicate mathematical ideas.

In the course of their negotiation, Nora and John invoked three aesthetic criteria; generativity, novelty, and visual appeal. These illustrate an already strongly developed sensitivity to the aesthetic qualities that can be associated with mathematical solutions—although perhaps not a conscious awareness of them. Through John's persuasion, Nora became curious about the magic number 9 as well and decided to turn her attention to a new investigation involving 9s. This is where her value-oriented epistemology emerged; she did not merely accept John's solution rather, she became convinced that his solution is interesting enough to warrant further investigation. That might well mirror one of the most important functions of the aesthetic in the day-to-day work of mathematicians, as one tries to convince another that some solution, result, or definition is interesting enough to warrant further work, for *both*.

I end this section by considering a noncomputer-based example, which is discussed by Dreyfus and Eisenberg (1986) also in the context of the mathematical aesthetic. They aimed to show that students lack the aesthetic sensibilities of mathematics by illustrating how students fail both to come up with "elegant" solutions and to appreciate those solutions once they are revealed. The problem was as follows: If you write all the numbers from 1 to 99,999, how many times will you write the number 7? I also tried this problem with students, but with a class of graduate mathematics education students, and under slightly different circumstances that I discuss shortly. Many students attacked the problem in a trudging sort of way, but first counting the 7s that would occur in each decade (for example, 1 between 30 and 39), taking care to count the extra seven that occurs between 61 and 70, and the many extra sevens that occur between 70 and 79. This repeated computation, with several adjustments, finally led them to a solution of 50,000. After arriving at their solution, several students tried to find more efficient ways, including counting the decades starting at 0 instead of 1. Several other students grew fed up with this plodding way of proceeding and devised other methods, often working together.

After 30 minutes of work, I offered the following solution, proposed by Dreyfus and Eisenberg: Write all the numbers as five digits abcde. If you write all the possible combinations of the letters, then every digit will occur equally often. There are 100,000 numbers with five digits, which means there are 500,000 digits in total. Because there are 10 digits between 0 and 9, each digit must occur exactly 50,000 times.

Contrary to Dreyfus and Eisenberg's students, who were interviewed individually, the students in this class expressed unanimous glee and preference for this clever solution. They repeated many of the arguments that

Dreyfus and Eisenberg offered for the solution's greater elegance, including its shortness, its generativity—in the sense that one can also answer the question of how many eights or twos one can write, its cleverness, and its perspicuity—the way it sheds light on itself, "explaining" why it holds true. These students were perhaps more appreciative and sensitive because of the context in which the problem-solving took place. The students had already been discussing and negotiating while working on the problem, so that the new solution wasn't a reflection on their weaknesses as individuals. They were also given enough time to reflect on and discuss their solutions, which motivated some to pursue "better" solutions, and likely put them in a frame of mind more conducive to aesthetic response than were Dreyfus and Eisenberg's students. Such experiences, which can abound in elementary number theory, help students move away from seeing mathematics simply as "a tricky way to find the answer."

SUMMARY

My primary goal is to show that elementary number theory offers many access points for mathematical appreciation, and engages many of the aesthetic sensibilities of learners that are productive, generative, and motivational in doing or engaging in mathematics. In most curricula, students go from grade 6 to university without seeing many ideas or methods of elementary number theory. The curriculum is certainly crowded enough nowadays, and adding more topics doesn't seem to be a wise way to proceed. However, what I have tried to do is challenge the kind of logic usually used in determining "the basics" in mathematics and show that students do not have to wait until graduate school mathematics to be able to appreciate the aesthetic dimension of mathematics—which is responsible for inspiring and sustaining the work of many, many mathematicians.

REFERENCES

Apostol, T. (2000). Irrationality of the square root of two: A geometric proof. *The American Mathematical Monthly, 107*(9), 241–242.

Aristotle. (1966). *Metaphysics* (H. Apostle, Trans.). Bloomington, IN: Indiana University Press.

Brown, S. (1973). Mathematics and humanistic themes: Sum considerations. *Educational Theory, 23*(3), 191–214.

Brown, S. (1993). Towards a pedagogy of confusion. In A. White (Ed.), *Essays in humanistic mathematics* (pp. 107–122). Washington, DC: Mathematical Association of America.

Brown, S., & Walter, M. (1983). *The art of problem posing*. Philadelphia, PA: Franklin Press.

Burton, L. (1999). The practices of mathematicians: What do they tell us about coming to know mathematics? *Educational Studies in Mathematics, 37*(2), 121–143.

Campbell, S., & Zazkis, R. (2002). *Learning and teaching number theory: Research in cognition and instruction*. Westport, CT: Ablex.

Devlin, K. (2000). *The math gene: How mathematical thinking evolved and why numbers are like gossip*. New York: Basic Books.

Dewey, J. (1934). *Art as experience*. New York: Perigree.

Dreyfus, T., & Eisenberg, T. (1986). On the aesthetics of mathematical thought. *For the Learning of Mathematics, 6*(1), 2–10.

Egan, K. (1990). *Romantic understanding: The development of rationality and imagination, ages 8–15*. New York: Routledge.

Ferrari, P. (2002). Understanding elementary number theory at the undergraduate level: A semiotic approach. In S. Campbell & R. Zazkis (Eds.), *Learning and teaching number theory: Research in cognition and instruction* (pp. 97–116). Westport, CT: Ablex.

Goldenberg, P. (1989). Seeing beauty in mathematics: Using fractal geometry to build a spirit of mathematical inquiry. *Journal of Mathematical Behavior, 8*, 169–204.

Gombrich, E. (1979). *The sense of order: A study in the psychology of decorative art*. Oxford, England: Phaidon Press.

Hardy, G. H. (1940). *A mathematician's apology*. Cambridge, England: Cambridge University Press.

Hawkins, D. (2000). *The roots of literacy*. Boulder, CO: University Press of Colorado.

Le Lionnais, F. (1986). La beauté en mathématiques [The beauty in mathematics]. In F. Le Lionnais (Ed.), *Les grands courants de la pensé mathématique* (pp. 437–465). Paris: Editions Rivages. (Original work published 1948)

Movshovits-Hadar, N. (1988). School mathematics theorems—An endless source of surprise. *For the Learning of Mathematics, 8*(3), 34–39.

Mumford, D., Series, C., & White, D. (2002). *Indra's Pearls: The vision of Felix Klein*. Cambridge, England: Cambridge University Press.

National Council of Teachers of Mathematics (NCTM). (1989). *Curriculum and evaluation standards for school mathematics*. Reston, VA: NCTM.

Papert, S. (1980). *Mindstorms*. New York: Basic Books.

Poincaré, H. (1956). Mathematical creation. In J. Newman (Ed.), *The world of mathematics* (pp. 2041–2050). New York: Simon & Schuster.

Rowland, T. (2002). Generic proofs in number theory. In S. Campbell & R. Zazkis (Eds.), *Learning and teaching number theory: Research in cognition and instruction* (pp. 157–184). Westport, CT: Ablex.

Russell, B. (1917). *Mysticism and logic*. New York: Doubleday.

Selden, A., & Selden, J. (2002). Reflections on mathematics education research questions in elementary number theory. In S. Campbell & R. Zazkis (Eds.), *Learning and teaching number theory: Research in cognition and instruction* (pp. 213–230). Westport, CT: Ablex.

Silver, E., & Metzger, W. (1989). Aesthetic influences on expert mathematical problem solving. In D. B. McLeod & V. M. Adams (Eds.), *Affect and mathematical problem solving* (pp. 59–74). New York: Springer-Verlag.

Sinclair, N. (2001). The aesthetic *is* relevant. *For the learning of mathematics, 21*(1), 25–33.

Sinclair, N. (2002). The kissing triangles: The aesthetics of mathematical discovery. *International Journal of Computers for Mathematics Learning, 7*(1), 45–63.

Sinclair, N. (2004). The roles of the aesthetic in mathematical inquiry. *Mathematical Thinking and Learning. 6*(3), 261–284.

Sinclair, N., Zazkis, R., & Liljedahl, P. (2004). Number Worlds: Visual and experimental access to elementary number theory concepts. *International Journal of Computers for Mathematics Learning, 8*(3), 235–263.

Stigler, J., & Hiebert, J. (1999). *The teaching gap: Best ideas from the world's teachers for improving education in the classroom*. New York: Free Press.

Tymoczko, T. (1993). Value judgements in mathematics: Can we treat mathematics as an art? In A. White (Ed.), *Essays in humanistic mathematics* (pp. 67–77). Washington, DC: The Mathematical Association of America.

Wells, D. (1990). Are these the most beautiful? *The Mathematical Intelligencer, 12*(3), 37–41.

Whitehead, A. N. (1929). *The aims of education*. New York: Macmillan.

Zazkis, R. (1999). Intuitive rules in number theory: Examples of the "the more of A, the more of B" rule implementation. *Educational Studies in Mathematics, 40,* 197–209.

Appendix A: The Number Worlds Applet

In what follows, I provide several examples of mathematical relationships inherent in Number Worlds.

Example 1: Stripes and Divisibility

The relationship between the input "show multiples of" and the grid width determines the type of pattern. For example, in the Natural World, showing multiples of 10 on a grid width of 10 results in one highlighted column (see Fig. 4.6) referred to as a "stripe." Showing multiples of 5 or 2 on the grid width of 10 results in two or five highlighted stripes, respectively. In general, a pattern of stripes emerges when the grid width is divisible by the input to "show multiples of."

Example 2: Stripes Shifted and What "Counts" as the First Multiple?

Showing multiples of 10 on a grid width of 10 forms a stripe; this stripe is shifted when the multiples are shifted. However, in order to have all the numbers in the stripe highlighted, a negative shift is required; the shifts of adding 2 and subtracting 8 in the Natural World are shown in Fig. 4.7. Of course, when working in the Whole World, which includes zero, this is no longer the case. Showing the multiples of 10, shift two would highlight the numbers 2, 12, 22, 32, and so forth.

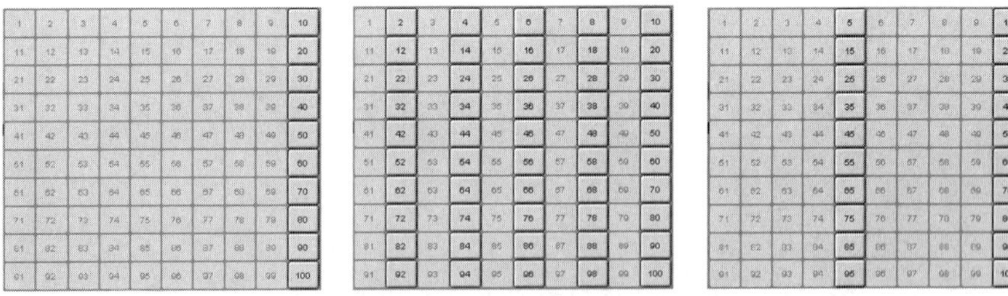

(a) Multiples of 10, gw = 10 (b) Multiples of 2, gw = 10 (c) Multiples of 5, gw = 10

FIG. 4.6. Showing multiples on a grid width of 10.

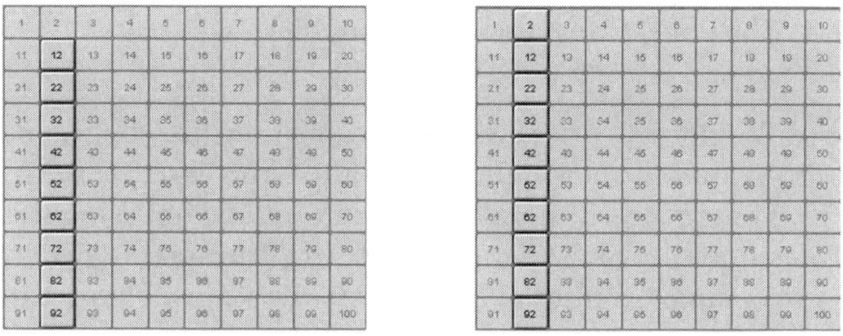

(a) Multiples of 10, shift 2 (b) Multiples of 10, shift –8

FIG. 4.7. Using the "shift" functionality.

Example 3: Diagonals and Nondivisibility

As shown earlier, a stripe pattern emerges when the input to multiples m is a factor of the grid width n. However, this is not the only way to produce a recognizable pattern. Figure 4.8 shows multiples of 2, 4, 5, and 7 on a grid width of 9. We refer to the patterns in (b) and (c) as diagonals, and refer to the pattern in (d) as disconnected diagonals. The pattern in (a) can be seen as diagonals as well; however, it is more naturally identified as a checkerboard pattern.

These images are examples of the following general relationship: "Diagonals" are displayed by multiples of m on a grid width of n, where $n \pm 1$ is divisible by m, and disconnected diagonals are displayed by multiples of m on a grid width of n, where neither n nor $n \pm 1$ is divisible by m. Checkerboards are a special case of diagonals and they are displayed by multiples of two on any odd grid width. (This precludes the creation of a real checkerboard pattern of width eight.)

By stacking the common differences on top of each other, the common-ness of the differences becomes visually obvious even when diagonals, rather than columns, are being displayed. We attend to this feature later, in our description of the research participants' work with the microworld.

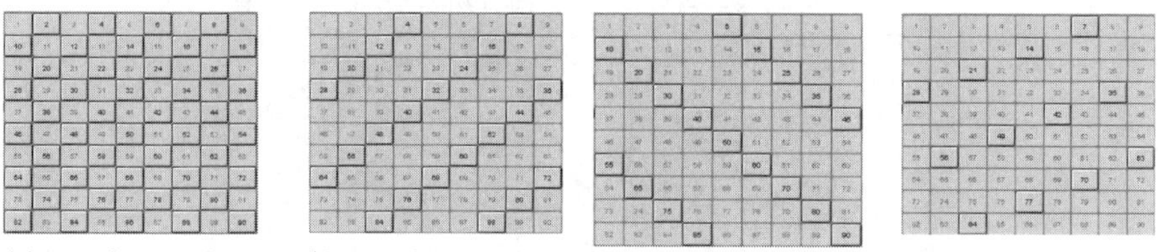

(a) m = 2, gw =9 (b) m = 4, gw = 9 (c) m = 5, gw = 9 (d) m = 7, gw = 9

FIG. 4.8. Relationships between multiples and grid widths.

Example 4: Shift and Congruence Classes

In examples 2 and 3, I describe how the relationships between the grid width and the input to show multiples of affect the visual patterns created on the grid. These patterns hold for any shift value; increasing the value of the shift will simply move the pattern over. To illustrate this, Fig. 4.9 invites the reader to compare multiples of 5, shift 0—as set by default—and multiples of 5, shift 3.

In general, by showing multiples with a shift, the user can generate any arithmetic sequence. The input for show multiples of corresponds to the common difference while the initial term is the sum of the inputs to show multiples and shift by in the Natural World. From a different mathematical point of view, the set of numbers highlighted by showing multiples of m with a shift of s actually forms a congruence class modulo s (using the convention that, for integers a and b, a is congruent to b modulo m if and only if m divides $a - b$; in other words, a and b have the same remainder in division by m).

Example 5: Prime Numbers

When prime numbers are highlighted on a grid of width 10 in the Natural World (see Fig. 4.10), one can *see* that the fourth, sixth, eighth, and 10th columns are empty—devoid of prime numbers. No highlighted cells appear in these columns because they contain only even numbers, which are divisible by 2 and therefore not prime. In addition, the fifth column contains only multiples of 5 (because the grid width is 10) and is thus similarly empty, except for the first cell. If the user selects Show Primes within the Odd World, the cells with the familiar prime numbers will be highlighted; however, all twin primes will appear in adjacent cells. Further, if the user selects Show Odds in the Prime World, every cell but the first will be highlighted (every prime number, except for 2, is odd).

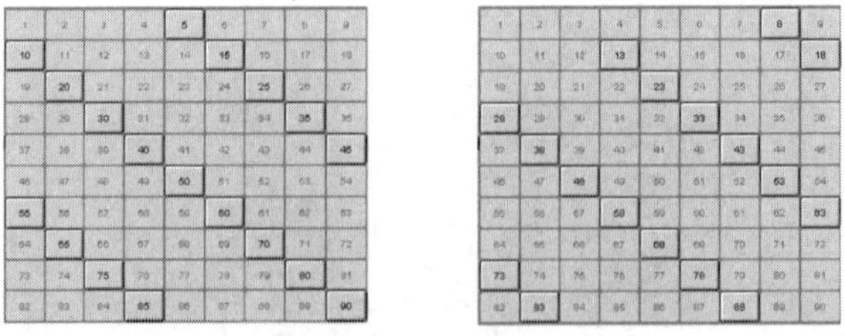

(a) m = 5, s = 0 (b) m = 5, s = 3

FIG. 4.9. Illustrating the effect of the shift on the pattern.

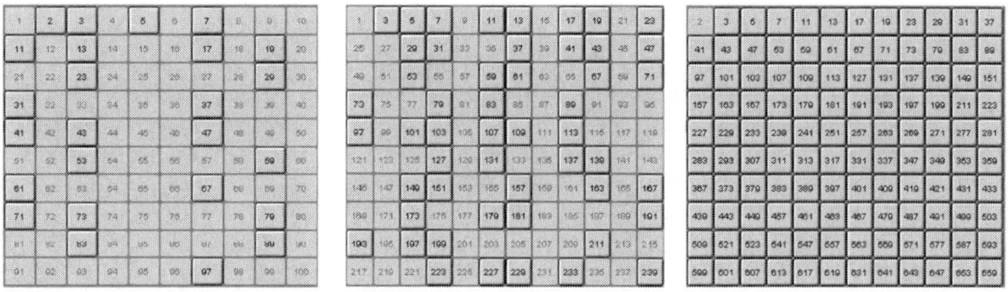

(a) Primes in Natural World (b) Primes in Odd World (c) Odds in Prime World

FIG. 4.10. Highlighting cells in different worlds.

Example 6: Arithmetic Operations and Closure

The operations of addition, subtraction, and multiplication function as they do on a regular calculator. However, the division operation returns the (whole number) quotient and the remainder, in order to preserve consistency with the number theory theme of the microworld. The arithmetic operations were included primarily for investigation of closure properties. If the result of an operation does not belong to the chosen number world, a message saying so appears above the grid. For example, the result of the operation 5 – 17 does not belong to the Natural World; the result of the operation 5 × 11 does not belong to the Prime World; the result of operation 3 + 9 does not belong to the Odd World.

Chapter 5

On the Path to Number Theory: Repeating Patterns as a Gateway

Rina Zazkis
Peter Liljedahl
Simon Fraser University

Prior research has described a variety of students' difficulties in understanding and implementing introductory concepts of number theory. Specific difficulties that were acknowledged included lack of connection between divisibility and division with remainder (Campbell, 2002; Zazkis & Liljedahl, 2002), lack of connection between factors and multiples (Zazkis, 2000) and lack of understanding of the distribution of numbers divisible by the same number among natural numbers (Ferrari, 2002; Zazkis & Campbell, 1996). In this chapter we put forward the following assertion: *Students' engagement with repeating patterns can help them in acquiring and strengthening concepts of elementary number theory and in establishing connections among theses concepts*. In particular we are concerned with the concepts of divisibility, factors, multiples, and division with remainder.

We start with a general discussion of patterns in mathematics and in mathematics education focusing explicitly on repeating patters. To outline the concepts of number theory that underlie repeating patterns, we zoom in on one particular problem situation and provide a mathematical analysis of possible approaches and of understandings that are required and that can be acquired in working on this problem. We then present empirical data from a group of preservice elementary school teachers dealing with the specific problem. In contrasting the work of these participants with the results from the similar population of students reported in prior research, we provide support for the assertion previously stated. We conclude with a suggestion for a pedagogical approach.

PATTERNS

Patterns in Mathematics Education Research and Practice

Mathematics is often perceived by the general public as the study of number. However, as Devlin (1994) suggested, this perception was accurate only up until about 500 B.C.E. Perceptions of what mathematics is have evolved through the centuries, undergoing many revisions. Words such as operations, interrelations, symbols, shape, structure, transformations, generalizations, abstraction, and combinations have made intermittent appearances in the dictionary definitions of mathematics. To capture the essence of mathematics and mathematical activity, while simultaneously embracing the variety of terms already mentioned, mathematics can be described as a science of patterns (Devlin, 1994). Moreover, the practitioner of this science, the "mathematician, like a painter or a poet, is a maker of patterns" (Hardy, 1940/1992, p. 84).

Mathematicians see in patterns not only structure, but also beauty; to them patterns have an aesthetic appeal. But unlike a painting or a poem, the mathematical pattern is a pattern of ideas, and as "ideas wear less with time than words" (Hardy, p. 85), the mathematical pattern has greater permanence. However, our concern is neither with the timelessness of ideas, nor with the specific ideas themselves, but with the use of patterns as a pedagogical tool for the delivery of the ideas that they represent.

Mathematics educators are in agreement about the contributions that patterns can make to the learning of mathematics. Mason (1996) saw patterns as a route to "expressing generality," a route that has become popular over the past 10 years within the school mathematics curriculum objectives in England (Orton & Orton, 1999). The NCTM identified "understanding patterns, relations and functions" as a crucial component of mathematics instruction at all grade levels (NCTM, 2000). Threlfall (1999) sees patterns as helpful in acquiring mathematical reasoning vital to learning, as a vehicle for working with symbols, as a context for generalization, as a conceptual stepping stone to algebra, and as a context for recognition, conjecturing, and communication of rules.

However, long before the aforementioned features are recognized, a pattern can be perceived by a young learner simply as a rhythm. The regularity of a pattern generates a rhythm that allows even young children to succeed in continuing a repeating pattern presented to them or in generating a repeating pattern of their own (Threlfall, 1999). Long before students are able to identify the rules or algebraic relationships that underlie the structure of a pattern, they are able to use the rhythm of a pattern to learn, to understand, and to create.

Patterns as a Cognitive Tool

Egan (1986, 1992, 1997) offered an insight into the seemingly innate ability of young children to work with patterns. Based on the work of Vygotsky, Egan formulated a theory of cognitive tools that emerge in learners as a result of their exposure to language—both spoken and written. One of these tools is an appreciation for, and ability to work with, rhythm and rhyme. This tool emerges at a very young age (2–4 years old) and not only coincides with, but is precipitated by, the acquisition of spoken language. On one hand, this theory explains the appeal that nursery rhymes and Doctor Seuss books have for such young minds; on the other hand, it explains why children develop such strong capabilities in working with auditory repeating patterns. The emergence of this tool, brought about by their exposure to language, creates an affinity for patterns at a deeply ingrained cognitive level.

"As rhythm is to the ear, so pattern is to the eye" (Egan, 1997, p. 243). Young children seem incapable of NOT learning how to continue and generate repeating patterns. Again, these abilities are deeply ingrained. As such, instruction should capitalize on students' cognitive tools, and, in particular, on their seemingly innate propensity toward patterns (Egan, 1997).

Are Patterns Undermined?

As dependent as mathematics is on patterns, and as powerful a cognitive tool pattern is, there exist two independent situations that conspire to undermine this "life blood" of mathematics and mathematical learning. The first of these is the minimal presence of any explicit treatment of patterns within curricular content. This may seem to be in contradiction with statements made earlier regarding to the promotion of patterns in NCTM and the English-equivalent curricular objectives. However, it should be noted that curriculum objectives and curriculum content are two very different entities. Whereas the curriculum objectives are often magnificent statements of all that is hoped for in a curriculum, its content is often reduced to a concise list of requirements to be mastered by students. Although it is commonly thought that the completion of curriculum content will result in the realization of its lofty objectives, this is often not the case. Patterns are a perfect example of this disparity between objectives and content. Although they are featured prominently in the objectives, or the rationale (as it is often referred to), they are often absent from the content requirements. As a result, patterns are often perceived as "enrichment" or "extracurricular" activities.

The second situation that is proving to be problematic for the promotion of patterns as a vital part of mathematical learning lies in the recent trend to emphasize the utility and applicability of mathematical knowledge in "real-life" situations. Consequently the structures and patterns of mathematical relationships are often deemphasized and implicit. Furthermore, the formalistic convention that dominates the practice of reporting results of professional mathematicians dictates that work be presented through the infallible logic of deduction. This protocol can draw the focus away from the rich patterns and relationships that support mathematical inventions.

All this being said, however, we would not want it to be thought that the teaching of mathematics is devoid of any emphasis given to patterns. The existing and the potential use of patterns in the teaching of mathematics for both its aesthetic and pedagogical qualities has been an area of some interest in the mathematics education research community.

REPEATING PATTERNS

Repeating Patterns in Mathematics Education Research and Practice

Mathematics education research discusses different kinds of patterns in mathematics; pictorial/geometric patterns, number patterns, linear patterns, patterns in computational procedures, and so forth. In this chapter, we focus on repeating patterns.

Repeating patterns are patterns with a recognizable repeating cycle of elements, referred to as a "unit of repeat" (Threlfall, 1999). For example, ABCABCABC ... is a repeating pattern, in which the unit of repeat has a length of three; 1,2,3,1,2,3,1,2,3, ... is a repeating pattern isomorphic to the previous example in which the attributes in the unit of repeat are numbers rather than letters (Liljedahl, 2004). Varying some attributes of elements (such as orientation, color, size, etc.), while keeping other attributes constant adds complexity to a repeating pattern. Repeating patterns can be created by the cyclic application of the unit of repeat, but they also appear naturally in mathematical structures. Consider for example the decimal expansion of rational numbers (4/11 = 0.3636363636 ...), the sequence of unit digits of any arithmetic sequence (**3**, **7**, 1**1**, 1**5**, 1**9**, 2**3**, 27, ...), or the sequence of last digits of consecutive powers of any natural numbers (3^n: **3**, **9**, 2**7**, 8**1**, 24**3**, ...).

Threlfall (1999) has focused on the use of one-dimensional repeating patterns in early primary years. Among his reasons for working with these structures is his belief that they develop a sense of sequencing and regularity. He has found that one way in which young children can succeed in generating repeating pattern or continue with a pattern presented to them is through a rhythmic approach. As we discussed earlier, this rhythmic approach is a "cognitive tool" on which instruction can capitalize. However, Threlfall also claimed that in order to generalize the pattern, a rhythmic approach is not sufficient; it is essential that the child develop a perception of the "unit of repeat" of the pattern. He further suggested that if work with repeating patterns is limited only to the primary years, the developmental immaturity of the child may prevent the acquisition of the concept of the unit of repeat. In order to gain the full benefit, exposure to repeating patterns should continue beyond the primary years, when meaningful generalization can be achieved.

We acknowledge the role that repeating patterns can play as a vehicle in working with symbols and generalization, and as such provide a powerful route toward algebra. We wish to extend this role further. Our claim is that repeating patterns provide a vehicle for directing student's attention to the multiplicative structure of natural numbers, and in such, provide a gateway to the introduction of the concepts of number theory. In what follows, we support this claim through the examination of a particular problem involving a repeating pattern and students' responses to this problem.

Repeating Pattern Problem

Consider the following problem:

Problem 1. *Imagine a toy train, in which the first car is red, the second is blue, the third is yellow, and again, the forth is red, the fifth is blue, the sixth is yellow, and the same pattern repeats for all the cars. What is the color of the 100th car? If the train has 100 cars, what is the number of the last yellow car?*

Even very young children can engage in the activity of continuing this pattern. They can form a train where each child stands for a car and declares or picks a label identifying its color. At an early age, the child's ability to continue the pattern can rely on recursive observations, that is, an ability to relate items to adjacent items (such as blue after red, yellow after blue, etc.), as well as on a "rhythmic" approach to memorizing the unit of repeat (that is, a chant of red-blue-yellow, red-blue-yellow, etc.; Liljedahl, 2004). However, as mentioned earlier, in order to approach the problem conceptually, it is essential to develop a perception of the unit of repeat in the repeating pattern (Threlfall,1999). Only then can one attend to the question "What is the color of the 100th car?"

This question may be too challenging for a very young child. Following Polya's tradition of problem solving (1985), we suggest a similar but simpler problem. What is the color of the 15th car? Whereas young children will play out the sequence explicitly, older ones may start paying attention to the "unit of repeat." Another way to simplify the problem is to consider a two-color train. This may present a wonderful, and for some learners the first, opportunity to consider even and odd numbers. Extending the unit of repeat, that is, the number of colors in a repeating pattern, can introduce or foster the concept of a multiple in elementary school years.

Increasing the numbers presents a similar but more difficult problem.

Problem 2. *Consider a 1,000 car toy train in a 7-color repeating pattern (red, orange, yellow, green, blue, purple, white). What is the color of the 800th car? What is the last blue car?*

In the next section, we consider the mathematical ideas that emerge in these problem situations.

Mathematical Analysis of a Repeating Pattern

The two problems presented in the previous section are examples of the following general structure:

> N different attributes are assigned to natural numbers in a cyclical repeating pattern of length N.[1] What attribute is assigned to a given number X?

In this section we analyze this structure using different mathematical tools, focusing on Problem 2.

1. Partition of Natural Numbers: Focusing on Remainders. Assigning a repeating pattern of attributes—in our case, colors—to natural numbers creates a structure that, in mathematical terms, is called a partition of the set. That is, according to the color attribute, a set of natural numbers is partitioned into disjoint subsets, such that every number belongs to one and only one subset. What can be said about each subset? Elements of each subset are congruent modulo n, where n is the length of a repeating pattern. This suggests that the remainder of a natural number in division by n uniquely determines the subset, and therefore the attribute, of this number. For example, the number 800 leaves a reminder of 2 in division by 7. Therefore, with the color-coding given earlier, the 800th car is assigned the same color as the second car, orange. Further, the question "what is the color of the last blue car in a 1,000 car train?" is equivalent to the question "what number is closest to 1,000 but smaller than 1,000 that leaves a remainder of 5 in division by 7?" This approach requires an overall structural consideration of the set of natural numbers and it may be preceded or complimented by a more sequential mathematical analysis of the situation at hand.

---

[1]This is not redundant; abcababcababcab ... exemplifies a repeating pattern with three attributes and unit of repeat of length five.

2. Repeating Unit: Focusing on Multiples. The correspondence between numbers and colors can be drawn sequentially, iterating through the colors in a cyclic fashion. A more efficient procedure is to consider a repeated application of the unit of repeat as one chunk (red-orange-yellow-green-blue-purple-white, red-orange-yellow-green-blue-purple-white, etc.) rather than the specific elements in this unit (red, orange, yellow, green, blue, purple, white, etc.). In an attempt to draw a connection between numbers and colors, we focus first on the last attribute of the unit of repeat. It is assigned to the number n, and then to 2n, 3n, 4n etc.; that is, the last attribute in the repeating pattern of a length n is assigned to the multiples of n. In our case, "white" is assigned to the 7th, 14th, 21st, 28th car and so on. That is, a car is white if and only if its position in the sequence is a multiple of 7.

From here we turn our attention to the cars that are colored red. Red is the first attribute in a repeating pattern and it is assigned to car positions of the form $7n + 1$, where $n \in W$. Similarly, for any k, $0 \leq k \leq 6$, k's attribute is assigned to car positions of the form $7n + k$. Focusing on multiples invites the following strategy: Every 7th car is white, therefore the 798th car is white, so the 799th is red, and subsequently the 800th car is orange. Similarly, we find that 994, which is a multiple of 7, is white and then we move through the sequence of colors to blue; 995 is red, 996 is orange, 997 is yellow, 998 is green and so 999 is blue.

3. Arithmetic Sequence: Focusing on Increments. Here we focus on numbers that have the same attribute, for example, the color yellow. From (1), all the yellow numbers leave a remainder of 3 in division by 7. From (2) all the yellow numbers are "3 counts up" from a multiple of 7. Furthermore, in considering the numbers with yellow attribute in order of appearance (3, 10, 17, 24, ...) we note that they form an arithmetic sequence with the common difference of 7. In general, elements of any subset in a partition modulo n, when considered in order, form an arithmetic sequence with a common difference of n.

It is evident that from a mathematical perspective, these approaches are equivalent. The difference is that they arise from shifting attention to different features of a repeating pattern. In (1) we consider the set of natural numbers as one entity, in (2) we focus on a repeating sequence of attributes, and in (3) we focus on each attribute with its corresponding numbers separately.

REPEATING PATTERNS IN STUDENTS' WORK

In what follows, we describe the themes that emerged in analyzing the solutions to Problem 2 of a group of preservice elementary school teachers:

Consider a 1,000 car toy train in a 7-colors repeating pattern (red, orange, yellow, green, blue, purple, white). What is the color of the 800th car? What is the last blue car?

We compare and contrast these themes with findings of prior research on the understanding of concepts and relations underlying elementary number theory. In particular, we focus on

1. the strategies employed by the participants in problem situations that present explicit versus implicit repeating pattern; and
2. the difficulties experienced by the participants in problem situations that present explicit repeating pattern versus difficulties reported in prior research on understanding elementary concepts in number theory.

The Main Strategies

Two main approaches were taken by students in considering the repeating pattern in the train problem. These approaches are exemplified in the responses of Cindy and Bill:

Cindy: $800 \div 7 = 114$ with a remainder of 2. So, the 800th car will have the same color as the 2nd car. Orange.

Bill: Every 7th car is white, so car 798 is white because $7 \times 114 = 798$. Then 799 is red and 800 is orange.

We refer to these two approaches as "division with remainder" and "counting from a multiple," respectively. Although the remainder strategy may appear as advantageous for a "mature" mathematical thinker, the strategy of "counting from a multiple" was preferred, that is, implemented by the majority of students. This is not surprising, because this strategy results mainly from a procedural rather than a structural consideration of the repeating pattern.

These same two strategies were also recognized in students' work with an arithmetic sequence (Zazkis & Liljedahl, 2002). In the following two excerpts, Megan and Lily consider the question of whether the number 704 is an element in an infinite arithmetic sequence 8, 15, 22, ...

Interviewer: And what about 704?

Megan: No, because that's got a remainder of (pause) 4, not 1 ... it needs to have a remainder of 1.

Interviewer: So can you please describe for me your general strategy? How would you decide whether a number I give you does belong to this sequence or doesn't belong to it?

Megan: Um, if it's divisible by 7, with the remainder of 1 then it does belong to the set.

Interviewer: Let's take one more. 8, 15, 22, 29 another sequence.
Lily: Okay, so this is a difference of 7 ...
Interviewer: How about the number 704?
Lily: (pause) 704, no (pause) this is, these numbers are plus 1 of multiples of 7, multiples of 7 plus 1 ...

Without expressing preference toward one strategy or the other, students were asked if they perceived the connection between the two. In the context of arithmetic sequence, even students who recognized, and confidently implemented both of the strategies were often unable to describe the connection, that is, to consider "remainder" as a distance from a multiple. Those students perceived the remainder as one of the numbers you get in performing division with remainder (Liljedahl & Zazkis, 2001). The following excerpt from the interview with Nina exemplifies such lack of connection.

Interviewer: Let's look at another sequence, 8, 15, 22, 29 ...
Nina: Okay, these go up by 7 ...
Interviewer: Now let's look at a big number, say 866. Do you think this number is an element in this sequence?
Nina: I have to think about it and calculate.
Interviewer: Fine. What would you do?
Nina: I'm just looking at the numbers, this 8 is just 7 + 1, and then 14 + 1, and then 21 + 1, so we have a 7 times table and one more.
Interviewer: Good. How does it help you?
Nina: I could just go on, like 29, 36, 43, so you see what I mean, so 6 you see could be the last digit, this doesn't help, but there could be a smarter way to calculate these [pause]
Interviewer: And if I told you that this number 866, when divided by 7 gives 123 and a remainder of 5. Will this help you?
Nina: Not sure. I can try timesing by 7 plus 1 and see what it could be.

Nina doesn't use the term multiple, or distance from a multiple, but she explicitly identifies the elements of the sequence using multiplication by 7 followed by the addition of 1. However, the explicit result of "remainder of 5" suggested by the interviewer does not guide her to decide whether the given number 866 is an element in the given sequence.

The lack of connection between "remainder" and "distance from a multiple" is a particular instance of a more general problem—the lack of connection between division with remainder and multiplication with addition. Campbell (2002) discussed this issue in detail. A very illustrative example is presented in a request to determine quotient and reminder in division by 6 of the number A, where A = 147 × 6 + 1. Fifteen out of 21 participants in Campbell's study calculated the dividend A and used a long division algorithm in order to answer this question.

Generalizing further, a fragile connection between multiplication and division with remainder could be seen as a manifestation of a "weak link" between multiplication and division in general. Consider for example the question of divisibility of M by 7, where $M = 3^3 \times 5^2 \times 7$. As reported by Zazkis and Campbell (1996), it was not uncommon to calculate the value of M and then divide it by 7 in order to conclude divisibility. Ferrari (2002) and Brown, Thomas, and Tolias (2002) confirmed this finding.

In the context of a problem that explicitly involved a repeating pattern, however, we note that the connection between the two strategies did not appear problematic. Although we engaged only a few participants in explicitly describing how the two strategies are related, we observed that in many cases participants applied the two strategies interchangeably, confidently changing between them or applying a combination of the two. Such a "combination" approach is exemplified in the following response:

> Dina: I know that 770 is divisible by 7, and so is 798, because you just added 28 to it. So 800 will be with reminder of 2. So this will have the second color which is red, orange, yes, orange.

In her approach, Dina does not perform the division with remainder, but she finds a remainder considering a close multiple. We recognize in Dina's approach her awareness of the fact that reminder and distance from a multiple are exactly the same.

Convenient Multiples

In applying the "counting from a multiple" strategy, it was common not to focus on the closest multiple, but rather on a "convenient" multiple.

> Carlo: 700 is white. So 707, 714, 721, ... um hm ... 770 is white, 777, 784, 791, 798 are white. So ... white, red, orange, 800 is orange.

In this example, Carlo starts "counting up" from 700. However, rather than immediately attending to 770 as the next possible benchmark for an "easy" multiple, Carlo starts a sequence of 707, 714, 721.... Then, he leaps to 770 and keeps counting up until 798 is eventually reached. We see in this strategy a strong preference toward addition, rather than multiplication. This can also be explained as an incomplete understanding of the fact that repeated addition of 7 is equivalent to adding a multiple of 7. The theme of additive dispositions has repeatedly appeared in prior research (e.g., Brown, Thomas, & Tolias, 2002). However, as shown in the next section, this additive perception may have advantages.

Multiples and "Every n^{th}" Property

Although Carlo's strategy of counting up from a familiar multiple may appear as lacking mathematical sophistication, it demonstrates an awareness that is vital in constructing an understanding of elementary number theory concepts; Carlo understands that every seventh car is white and also that white-colored cars are numbered by multiples of 7. This understanding encompasses an important underlying idea of the number structure, the idea that "every seventh number is divisible by 7," or in general, that "every n^{th} number is divisible by n."

Prior research has shown that this property is not among the properties that students take for granted. For example, participants in a study of Zazkis and Campbell (1996) had difficulty in deciding whether there is a number divisible by 7 in a given interval of 10 (they were asked to consider numbers between 12,358 and 12,368) without explicitly finding such a number in this interval. Furthermore, after correct calculation of remainder in division of 12,358 by 7, not everyone could determine, without calculation, what would be the remainder in division by 7 of 12,359. Ferrari (2002) reported a similar finding. Participants in his study were engineering and computer science majors who were asked, "Is there an integer x such that $M \leq x \leq M + 10$ and 8 is a divisor of x?" The number M was previously given to the participants as $M = 3^4 \times 5^3 \times 7^6 \times 19^8$. This question appeared to cause difficulty to the participants as they paid explicit attention to the value of M in their attempts to deal with the question.

In the context of a repeating pattern, however, this didn't seem to be an issue. During the interviews regarding the train question, students were presented with the question: Is there a yellow car in the first 10 cars after 200? Not only was this question always answered correctly, it was done so with an air of triviality. That is, the students viewed the result as being obvious.

Every n^{th} Property—Extended

In mathematics the idea that "every n^{th} number is divisible by n" is naturally extended to the partition of natural numbers into disjoint sets by the relation of congruency modulo n (that is, elements in each set of such partition are congruent to each other modulo n). This idea is expressed in terms of our pattern as "every seventh car is white." However, depending on where the count begins, it is equally true that every seventh car is blue and every seventh car is orange as demonstrated by Stacey's comments.

> Stacey: Since the 800th car is orange then ... 807 will also be orange, and so will 814, 821, 828.

Stacey is initially invoking this property in an attempt to get from the 800th car to 1,000th car. She eventually changes her strategy to one of counting up from a multiple when she realizes that her choice of approach is too time consuming.

Although it may appear as "trivial" when considering train cars, grasping the formal equivalent of this property—that, for example, numbers leaving a reminder of 5 in division by 7 are "7 apart" on the number line—has proven to be difficult. The idea of partitioning of natural numbers was explored in a study that focused on arithmetic sequences (Zazkis & Liljedahl, 2002). In one of the interview questions, students were asked to consider whether a given (large) number was an element of a given arithmetic sequence. Our analysis showed a distinction between sequences of multiples (e.g., 7, 14, 21, 28, ...) and sequences of so-called "nonmultiples" (e.g., 8, 15, 22, 29, ...). Although the divisibility of a number by *d* (*d* is the common difference) gave a clear indication of belonging to a sequence of multiples, the lack of divisibility by *d* left students uncertain of its membership in a sequence of "nonmultiples." In the following excerpt, Leah is first considering the arithmetic sequence 7,14, 21, 28 ...

> Interviewer: Please look at the arithmetic sequence 17,34, 51, 68 and so on. I would like to ask you about the number 204, is it an element in this sequence?
> Leah: If it is a multiple of 17, it is.
> Interviewer: And if not?
> Leah: Then it wouldn't be. But this one is. 204 is 12 times 17.
> Interviewer: Can you please give me an example of a large number which is in this sequence and a large number which is not.
> Leah: 170, 340, 1700
> Interviewer: And how about a number that is not in the sequence?
> Leah: Just take anything without this factor, with no 17 as a factor.

Leah identifies the given sequence as a sequence of multiples of 17 and therefore the elements of the sequence are multiples of 17 whereas "anything with no 17 as a factor" is not an element in this sequence. However, the problem appears when this reasonable strategy for distinction between elements and nonelements in case of a sequence of multiples is mirrored, rather than adjusted, for the sequence of nonmultiples in question.

Interviewer: Let us look at a different sequence, say 8, 15, 22, 29, and it is an arithmetic sequence ...
Interviewer: Okay, so can you give me an example of a number that you believe is not in the sequence and an example of a number that you believe is, or could be in the sequence?
Leah: Um, I don't think 714 would be in the sequence, um, a number that could be, I would just pick a number that hasn't a factor of 7, so like 511 possibly, or something
Interviewer: And you are saying possibly because ...
Leah: Just, I just picked a number that wasn't, didn't have 7 as a factor.

From Leah's response it is clear that she recognizes the given sequence as the sequence of nonmultiples. Therefore a number that has 7 as a factor cannot be an element in this sequence. However, Leah fails to recognize the structure in the sequence of multiples. For her, the only feature of the elements of this sequence is lack of divisibility by 7. As such, any number that does not have a factor of 7 could be an element in this sequence. Leah's approach, common among her classmates, is an indication that extending the property of "every nth" from divisibility to division with remainder, or from multiples to "multiples plus" cannot be taken for granted.

It may appear surprising that properties of numbers that are easily detected by students in one problem situation are often overlooked in a different problem situation. This is especially true with respect to properties that the mathematically inclined tend to take for granted, "every nth" being just one such example. Although this property emerged naturally in considering the colors of cars, it was not utilized in the context of an arithmetic sequence. Further research is needed to gain a better understanding of the degree to which exposure to concepts of elementary number theory through repeating patterns can help students in working with these concepts in other problem situations.

Every nth Property—Misgeneralized

The accumulation of our research into preservice elementary school teachers' understanding of elementary number theory has consistently shown that the inherent relationships of "every nth" is difficult for this population to master. Nowhere is this more apparent than in the students' misgeneralizations of this property. Although infrequent occurrences were noted, the use of repeating patterns did not inoculate the students from mistakes. In this section, we present some of these misgeneralizations.

- A notable misapplication of the above mentioned property "every 7th number is divisible by 7" appeared when students assigned to a car, which number was divisible by 7, the color "red," the color of the first car, rather than "white"—the color of the seventh car. These students noted the property of "every seventh," but applied it starting with the first, rather than the seventh, element (Liljedahl & Zazkis, 2001).
- Another improper generalization repeatedly appeared when participants generated the property of the kth attribute to "every k^{th}" car, without paying attention to the unit of repeat. Claims such as "Because the blue car is in the 5th position, all the multiples of 5 will be blue" illustrates this approach.
- For a number of students that clearly and correctly identified that multiples of 7 are the key to solving the pattern, the ways to determine such multiples were improperly generalized. We witnessed claims that "the 137th car is white because 137 is divisible by 7 as it ends in 7" and also "151 is divisible by 7 because of the sum of the digits"—which are based on overgeneralization of familiar divisibility rules for 5 and 3. Although these claims were infrequent, they confirmed the appearance of similar improper generalizations reported in prior research (Zazkis & Campbell, 1996).

Overview

In considering students' work with problems involving repeating patterns and comparing this to students' work with other number theory related problems, we have shown that:

1. similar strategies emerge in students' approaches to problem solving
2. many issues of difficulty reported in prior research are either resolved or refined in consideration of repeating patterns.

In (1) we focused on consideration of remainder in division and a distance from a multiple as links between multiplication and division in general and multiplication and division with remainder in particular. Although both strategies were evident in participants' approaches, it appeared that in considering repeating patterns, these strategies were better connected.

In (2) we focused on the distribution of multiples of n and the distribution of numbers in the same equivalence class modulo n. Problematic issues presented in prior research appeared of lesser concern; that is, the difficulties in understanding the distribution were individual occurrences rather than common trends. We do not suggest that the "transfer" of understanding of multiplicative structures between different problem situations is effortless. But we do suggest that consideration of repeating patterns provides a helpful pedagogical tool for building this understanding.

CONCLUSION

Repeating patterns are not just for the nursery school. Nor is competence with repeating patterns merely a cognitive tool (albeit a powerful one). In this chapter, we present the equally powerful idea that repeating patterns can be used as a pedagogical tool to help facilitate the understanding of elementary number theory concepts for students long after they leave the nursery.

There is a recent trend in mathematics education to make mathematics "relevant" to the students by presenting it in "context." Unfortunately, "relevance" and "context" are often interpreted as activities related to "everyday" or "real" life. We agree with Nemirovsky (1996) that "real contexts are to be found in the experience of the problem solvers" (p. 313), rather than in the formulation of the problem. Numbers themselves offer a "context" for investigation and for a rich background of ideas and experiences.

The idea that patterns can aid in the acquisition of mathematical concepts is not new. Focusing on patterns is advocated in the research literature as a stepping stone in the generalization approach to algebra (Lee, 1996). In particular, Threlfall (1999) argued that linear, or one-dimensional, repeating patterns represent a path toward number patterns in algebra. Extending these claims, we suggest that attending to number patterns is also a gateway to number theory. We believe that consideration of repeating patterns can both introduce and enhance concepts and relationships underlying elementary number theory, and especially the multiplicative structure of natural numbers. Furthermore, a learner's propensity for patterns could make the transition natural and accessible.

REFERENCES

Brown, A., Thomas, K., & Tolias, G. (2002). Conceptions of divisibility: Success and understanding. In S. R. Campbell & R. Zazkis (Eds.), *Learning and teaching number theory: Research in cognition and instruction* (pp. 41–82). Westport, CT: Ablex.

Campbell, S. R. (2002). Coming to terms with division: Preservice teachers' understanding. In S. R. Campbell & R. Zazkis (Eds.), *Learning and teaching number theory: Research in cognition and instruction* (pp. 15–40). Westport, CT: Ablex.

Devlin, K. (1994). *Mathematics: The science of patterns*. New York: Scientific American Library.

Egan, K. (1986). *Teaching as story telling*. London, Ontario: The Althouse Press.

Egan, K. (1992). *Imagination in teaching and learning*. London, Ontario: The Althouse Press.

Egan, K. (1997). *The educated mind*. Chicago: The University of Chicago Press.

Ferrari, P. L. (2002). Understanding elementary number theory at the undergraduate level: A semiotic approach. In S. Campbell & R. Zazkis (Eds.), *Learning and teaching number theory: Research in cognition and instruction* (pp. 97–116). Westport, CT: Ablex.

Hardy, G. H. (1992). *A mathematician's apology*. Cambridge, England: Cambridge University Press. (Original work published 1940)

Lee, L. (1996). An initiation into algebraic culture through generalization activities. In N. Bednarz, C. Kieran, & L. Lee (Eds.), *Approaches to algebra* (pp. 87–106). Dordrecht, Netherlands: Kluwer.

Liljedahl, P. (2004). Repeating pattern or number pattern: The distinction is blurred. *Focus on Learning Problems in Mathematics, 26*(3), 24–42.

Liljedahl, P., & Zazkis, R. (2001, July). Analogy in the exploration of repeating patterns. *Proceedings of the International Conference for Psychology of Mathematics Education* (pp. 305–312). Utrecht, Netherlands: Utrecht University Press.

Mason, J. (1996). Expressing generality and roots of algebra. In N. Bednarz, C. Kieran, & L. Lee (Eds.), *Approaches to algebra* (pp. 65–86). Dordrecht, Netherlands: Kluwer.

National Council of Teachers of Mathematics (NCTM). (2000). *Principles and standards for school mathematics.* Reston, VA: Author.

Nemirovsky, R. (1996). A functional approach to algebra: Two issues that emerge. In N. Bednarz, C. Kieran, & L. Lee (Eds.), *Approaches to algebra* (pp. 295–313). Dordrecht, Netherlands: Kluwer.

Orton, A., & Orton, J. (1999). Pattern and the approach to algebra. In A. Orton (Ed.), *Pattern in the teaching and learning of mathematics* (pp. 104–120). London: Cassell.

Polya, G. (1985). *How to solve it* (2nd ed.). Princeton, NJ: Princeton University Press.

Threlfall, J. (1999). Repeating patterns in the primary years. In A. Orton (Ed.), *Pattern in the teaching and learning of mathematics* (pp. 18–30). London: Cassell.

Zazkis, R., & Campbell, S. R. (1996). Divisibility and multiplicative structure of natural numbers: Preservice teachers' understanding. *Journal for Research in Mathematics Education, 27*(5), 540–563.

Zazkis, R. (2000). Factors, divisors and multiples: Exploring the web of students' connections. *Research in Collegiate Mathematics Education, 4,* 210–238.

Zazkis, R., & Liljedahl, P. (2002). Arithmetic sequence as a bridge among conceptual fields. *Canadian Journal of Science, Mathematics and Technology Education, 2*(1), 93–120.

Chapter 6

Learning by Teaching: The Case of Sieve of Eratosthenes and One Elementary School Teacher

Roza Leikin
University of Haifa

> *… it is only when you come to teach something that you really come to understand it.*
>
> —Mason (1998, p. 260)

This research analyzes the development of teachers' knowledge as evidenced in the case of one elementary school teacher (Nurit) who was asked to teach a lesson on prime numbers using the Sieve of Eratosthenes. This study is based on the assumption that teachers learn when teaching and it reflects complexity of teachers' knowledge in the 3-D model, which is suggested here as a framework for the analysis of learning through teaching. It is aimed at disclosing the mechanisms of this learning and the relationships between three dimensions of teachers' knowledge; its kind, its forms, and its sources. The chapter presents several accounts that describe the role that different forms of knowledge play in the development of teachers' knowledge through teaching. The episodes in each account indicate the mechanisms of knowledge development.

RATIONALE

Members of the research community very often address the fact that teachers learn from teaching experience (e.g., Cobb & McClain, 2001; Kennedy, 2002; Lampert & Ball, 1999; Lesh & Kelly, 1994; Ma, 1999; Mason, 1998; Shulman, 1986; Wilson, Shulman, & Richert, 1987). Comparison between experienced and novice teachers (Berliner, 1987; Leinhardt, 1993) as well as the development of interns' knowledge (Roth, 1998) attests to the learning-through-teaching phenomenon. Although experience and expertise are not equivalent terms, teachers' expertise is usually considered a function of their experience. For example, according to Wilson, Shulman, and Richert, (1987), teachers' pedagogical reasoning begins with comprehension that consists of critical understanding of a set of ideas, a piece of content, in terms of both its substantive and its syntactic structure. Furthermore, teachers are involved in a transformation process associated with teachers' planning and designing activities, instructional activities, evaluation, and reflection. As a result, teachers attain new comprehension, which is enriched by fresh understanding enhanced by increased awareness of the purposes of instruction, the subject matter of instruction, and the participants (teachers and students).

The key features of the learning-through-teaching (LTT) process, as noted in the research literature, are teacher interaction with students and instructional materials, as well as their teacher reflection (e.g., Dewey, 1933; Schön, 1983). Several cyclical models of teaching included these components of teacher instructional activity emphasizing the fact that teacher knowledge develops through instructional practice as a result of teachers' interactions with their students. Such a view of teaching (e.g., Artzt & Armour-Thomas, 2002; Simon, 1997; Steinbring, 1998) makes it reasonable to expect development in teacher knowledge by analyzing this interactive process. Whereas students are supposed to learn deliberately and the teachers' main purpose is to support students' learning, the teachers themselves learn through teaching unintentionally (indirectly in terms of Stigler & Hiebert, 1999).

Leikin (2005) suggested a model of teachers' instructional interactions that allow detailed analysis of teachers' interactions in a system of six qualities: (1) the *purpose* for which a teacher may interact with students; (2) *initiation* of interaction by the teacher or by the pupils; (3) *motives* for interacting, which may be external if they are prescribed by the given educational system, or internal, being mostly psychological, including cognitive conflict, uncertainty, disagreement, or curiosity; (4) *reflection* on or in teachers' or students' previous experiences; (5) *actions* that support the interactive process; and (6) *focus* of interaction, which may be mathematical or pedagogical. Each of these six qualities may be divided into large categories so that any interaction may be seen as a vector constructed of six coordinates. This model of the teacher–students interactions helps in explaining mechanisms of teacher learning through teaching (Leikin, 2004, 2005) as well as in analyzing teachers' proficiency in managing the student-centered class.

Although broadly addressed, the learning-through-teaching phenomenon has not been examined carefully with regard to the development of different types of teachers' knowledge. The main body of research on learning-through-teaching mainly focuses on teachers' continual inquiry into students' thinking and learning (e.g., Franke, Carpenter, Levi, & Fennema, 2001; Kennedy, 2002; Schifter, 1998). In some studies, researchers who are expert teachers analyze their own pedagogical growth (Chazan, 2000; Lampert, 2001). In other cases, research on learning-through-teaching focuses on a microteaching situation (Leikin, Berman, & Zaslavsky, 2000). So, all in all, the educational community has relatively limited understanding of what changes in different types of teachers' knowledge, or how those changes come about, especially in the field of mathematical knowledge in an authentic classroom situation.

THEORETICAL BACKGROUND: TEACHERS' KNOWLEDGE

Epistemological analysis of teachers' knowledge reveals significant complexities in its structure (e.g., Scheffler, 1965; Shulman, 1986; Wilson et al., 1987). Scheffler (1965) argued for the importance of epistemological analysis as logical analysis of knowledge as a complex concept. He suggested a detailed analysis of knowledge in the educational context, considered relationships between knowledge and teaching as well as between knowledge and learning.

To address the development of teachers' knowledge by teaching, I first analyze the complexity of teachers' knowledge and present different approaches to the classification of knowledge. I suggest compounding these classifications into a 3D model of the teacher's knowledge structure (Fig. 6.1).

Dimension 1: Kinds of Teachers' Knowledge

Shulman (1986) suggested three main components that determine different kinds of teachers' knowledge: Subject matter knowledge, pedagogical content knowledge, and curricular content knowledge. Teachers' subject-matter knowledge includes knowledge of mathematical concepts, different representations of those concepts and connections between them, knowledge of connections between different mathematical concepts and their implementation for solving mathematical problems, and knowledge of the history and philosophy of mathematics. Teachers' pedagogical content knowledge includes knowledge of students' cognitive processes associ-

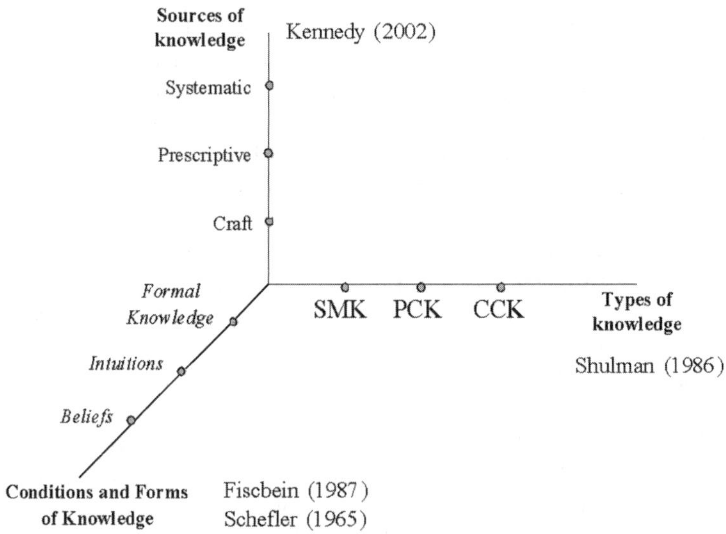

FIG. 6.1. Three dimensions of teachers' knowledge.

ated with particular mathematical topics and with particular learning skills, teachers' capacity to adapt learning materials to the students' needs and abilities, knowledge of different learning settings, and competence in connecting a particular learning setting with a particular mathematical topic or a particular mathematical task. Teachers' curricular content knowledge includes knowledge of different types of curricula, understanding different approaches to school mathematics, awareness of different learning sequences within a particular curriculum, and the authority to make choices. These categories are not disparate; they are mutually influential and complementary. In the analysis I address data that is explicitly and clearly related to one of these categories, as well as also consider mutual relationships among different kinds of teachers' knowledge.

Dimension 2: Sources of Teachers' Knowledge

Kennedy (2002) classified teachers' knowledge according to the sources of its development. The first type is teachers' craft knowledge, which is largely developed through experience and tends to be atheoretical and idiosyncratic. Kennedy argued that teacher knowledge of this type is mainly intuitive and makes an incremental and cumulative impact on teachers. According to that study, craft knowledge is mainly associated with teachers' concerns regarding orderly task flow, and students' willingness to participate in classroom activities. The other type of knowledge, according to Kennedy (2002), is systematic. Knowledge of this type is acquired mainly through systematic studies in colleges and universities, and through reading research articles, journals, and professional books. This knowledge tends to be theoretical, codified, and abstract, and concerns teachers' sense of responsibility. The last type of knowledge in this classification is prescriptive. This is acquired through institutional policies, which are transparent in tests, accountability systems, and texts of diverse nature. It tends to consist of "should and "ought to" statements. Knowledge of this type is more than craft knowledge and less theoretical than systematic knowledge. It is motivated mainly by teachers' sense of responsibility to students and to community. Teachers' prescriptive knowledge tends to be transient as the policy changes.

As this study focuses on the learning-through-teaching phenomena it mainly examines the mechanism for the development of teachers' craft knowledge. Whereas Kennedy (2002) argued that craft knowledge is mainly associated with teachers' pedagogical content knowledge (orderly task flow and students' willingness to participate), this study is interested in connections between craft knowledge and different types of teachers' knowledge with special attention to teachers' mathematical knowledge.

Dimension 3: Forms and Conditions of Knowledge

Kennedy (2002) also posited a correspondence between different sources of knowledge and different forms of knowledge, (e.g., craft knowledge is mainly intuitive, systematic knowledge is mainly formal). This research analyzes connections between different forms of knowledge and different kinds of teachers' knowledge as they develop through teaching practice. It differentiates among teachers' intuitive knowledge as determining teachers' actions that cannot be premeditated (e.g., Atkinson & Claxton, 2000), their formal knowledge that is mostly connected to planned teachers' actions, and teachers' beliefs that are expressed in teachers' conceptions of teaching.

Knowledge and Belief. Regarding conditions of knowledge, Scheffler (1965) distinguished between knowing and believing. Knowing has "propositional and procedural nature" whereas believing is "construable as solely propositional" (p. 15).

Cooney (2001) continued Scheffler's definition of beliefs as "abstract things, in the nature of a habit or readiness" that express "a disposition to act in a certain way under certain circumstances" (Scheffler, 1965, p. 76), with an additional condition. For Cooney, verbal proclamations of beliefs must be followed by actions consistent with and proving those beliefs. In the educational context, learning precedes believing and knowing.

According to Scheffler, there are three conditions for knowing that:

> X knows that Q if and only if (i) X believes that Q, (ii) X has adequate evidence that Q, and (iii) Q [is true].[1] (p. 21)

From this definition, it follows that believing is one of the conditions of knowing, implying that if somebody knows that Q he believes that Q. However, believing is not sufficient for knowing. Scheffler suggested that knowing is a term of wider range than believing. One may "know that" and "know how to" whereas only "believe that" is possible and "believe how to" does not make sense.

Formal and Intuitive Knowledge. Fischbein (1984) in his analysis of mathematics as human activity considered three main components; the formal, the algorithmic, and the intuitive. The formal component in Fischbein's categorization addresses the "knowing that" condition of knowledge, and its algorithmic component of human mathematical activity clearly correlates with knowing how and being able to proceed in a certain situation. An intuition is a form of knowledge (Fischbein, 1982) distinct from formal, that is, deductively based and inductively based knowledge in its intrinsic character: No direct need is felt to prove formally or factually the respective interpretation or representation.

Atkinson and Claxton (2000) argued for the intuitive nature of teachers' knowledge in action. They suggested that teachers' intuitions are an important part of teachers' professionalism:

> It is self-evident that much of what teachers do, in the heat of the moment, it is not premeditated; it is intuitive. A situation arises; the teacher responds, and only later, if at all, will she or he pause to figure out what was going on, and why they did what they did. (p. 1)

Just as Fischbein argued for intuitive cognition as characterized by being self-evident, I suggest to define intuitions by the partial set of conditions presented by Schefler for formal knowledge. Specifically, I propose that intuitive knowledge (X intuitively knows that Q) fulfils two of three conditions: X believes that Q, and X thinks that there is adequate evidence that Q. Then, intuitions are immediate mental structures that stand between knowledge and belief. In the cases in which intuitive knowledge "Q" is supported by deductive proof "Q is true" it may be transformed into the category of formal knowledge, whereas in other cases, intuitions may be stored in the form of belief (see Fig. 6.2).

---

[1] My addition.

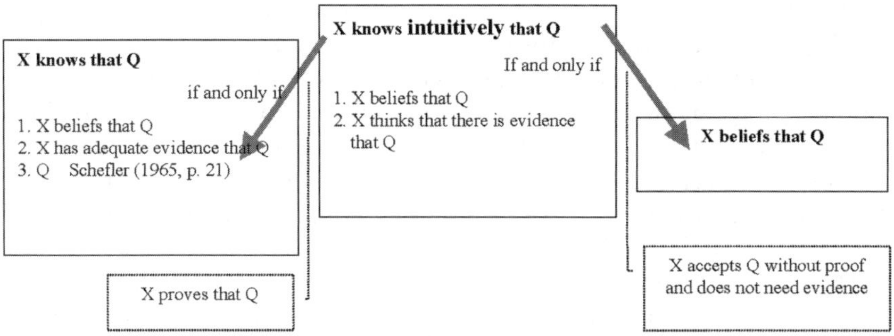

FIG. 6.2. Relationship between formal knowledge, intuitive knowledge, and beliefs.

Whereas Fischbein comprehensively analyzed intuitions in the field of subject matter, this article attempts to consider teachers' mathematical and pedagogical intuitions alike as a source for the development of teachers' mathematical and pedagogical craft knowledge. It differentiates teachers' formal knowledge, from teachers' intuitive knowledge as insights in the form of immediate mental structures in the field of teachers' content knowledge. Note, that In this research formal knowledge refers to what Fischbein (1984) calls either formal or algorithmic knowledge. An additional distinction is made here between teachers' knowledge (intuitive and formal) and teachers' beliefs, which usually appear in the form of opinions that express teachers' dispositions towards mathematics and pedagogy.

THE STUDY

This chapter presents one case from an extensive study analyzing multiple cases of teachers' learning through teaching students. The methodology followed a qualitative research paradigm (Strauss & Corbin, 1990). Different cases were collected simultaneously, and complemented each other in an inductive process while the research interpretations and findings, crystallized as pieces of information, accumulated. The analysis was done with respect to the categories of knowledge presented in the 3D scheme.

The Aim and the Questions

Three dimensions of teachers' knowledge serve as a framework for the whole study, which outlines relationships between the different dimensions in the process of the development of knowledge through teaching (see Fig. 6.1). It explores the development of teachers' mathematical and pedagogical content knowledge through teaching practice and addresses the relationship between different forms and sources of knowledge. The whole study attempts to answer three research questions: (1) How does teachers' content knowledge develop in the course of teaching a neglected (by a teacher) mathematical topic? (2) What role do different forms of knowledge play in the development of teachers' knowledge through teaching? (3) What roles do different sources of knowledge play in the development of teachers' mathematical knowledge?

This chapter focuses on one specific case from the multiple-case study. It places specific emphasis on the development of teachers' craft knowledge as knowledge developed through teaching, and addresses the first and the second research questions.

DATA COLLECTION AND ANALYSIS

Overall, 29 teachers participated in the whole study on a short-time basis. Of these, 7 teachers were elementary school teachers and 22 were secondary school teachers. The teachers' experience varied from 3 to 17 years. Each of them in this group taught a mathematical task given by the researcher that was identified as unfamiliar or neglected. The classes varied from fourth grade to 10th grade and from 25 to 39 students in a class. The duration of teaching varied from one to three lessons depending on the students' pace of learning. In most cases, the mathematical tasks were related to the curriculum that the teachers taught at that period of time but they were unfamiliar to the teachers. The use of unfamiliar mathematical tasks in this study made it possible to intensify the haziness of the situations in which teachers learned mathematics when teaching their students. All the data were videotaped and transcribed. Data analysis of any piece of information was performed by an interviewer and the present author. All the teachers participated in the study voluntarily.

I was aware that the development of mathematical understanding is a longitudinal process, and it would be unreasonable to expect that teachers develop their understanding either procedurally or conceptually (Skemp, 1987) in one or two particular lessons. However, as an indication of teachers' learning, I considered pieces of information that presented teachers' mathematical noticing (on this concept see Mason, 2002), expressed, for example, in changes in their mathematical terminology, changes in their ways of defining mathematical concepts, "picking" new mathematical ideas or new ways of solving problems from the students, or coming up with new ideas for teaching purposes.

In all the cases the data were collected in triads: The teachers were asked to participate in a prelesson interview when planning the lesson, to allow the researchers to record their lessons, and to participate in the postlesson interview when analyzing the lesson. Teachers' handwritten notes during the interviews were collected.

Overall, the study is based on a collection of cases focusing on different teachers, on different grades, and on different mathematical topics. The cases also differed in the amount of episodes that included teachers' learning. In this chapter, I describe one particular case focused on teaching prime numbers with the Sieve of Eratosthenes in elementary school. This case partly answers the research questions of the whole study, illustrates several regularities in teachers' learning through teaching found in the whole study and helps in explaining the methodology employed. In some cases collected in the larger study, the teachers learned more when planning the lesson; in other cases, teachers learn during the lessons, and sometimes when analyzing a lesson. The particular case in this chapter represents mainly the teacher's learning when planning the lesson.

ONE CASE STUDY: NURIT, PRIME NUMBERS AND SIEVE OF ERATOSTHENES

Nurit—The Teacher

Nurit (pseudonym) was a primary school teacher with 7 years of teaching experience after completing BEd in elementary school teaching with specialization in mathematics. She had taught mathematics in all the grades (one to six) at elementary school, and participated in this research experiment with her fourth-grade students. We gave her a whole set of instructional materials (worksheets and Web sites) for teaching the topic "prime and composite numbers" from which she was asked to choose materials she considered suitable for her lessons and that matched her personal preferences. Among other materials, she chose the Sieve of Eratosthenes (see Fig. 6.3) to introduce prime numbers to her students.

Mathematical Context: Primes, Composites and Sieve of Eratosthenes

This topic offers opportunities for enriching students' mathematical language, making connections with other related topics like prime decomposition and divisibility, discussing the historical development of mathematics, exploring the Internet, and gaining further understanding of mathematics topics already covered (e.g., Yolles, 2001). The Sieve of Eratosthenes is one of the ways of developing students' understanding of the concept of prime numbers. The students can be asked to generate prime numbers using this tool and to predict the number of primes that we will find on the sieve activity sheet. It also allows finding patterns of multiples. Using the sieve activity

Sieve of Eratosthenes

1	2	3	4	5	6
7	8				

Fill the table by all whole numbers from 1 to 200 in order.

Perform the following:

a. Cross 1
b. Circle 2, cross all other numbers in the columns, of 2, 4 and 6.
How can you characterize these numbers? Complete the sentence:
The crossed numbers are ____________________

c. Circle 3, cross all the multiples of 3

d. Circle 5, cross all the multiples of 5

e. Circle 7, cross all the multiples of 7

f. Circle next (after 7) uncrossed number, cross all the multiples of this number

Continue circling and crossing the numbers.

All the circled numbers are called ***prime numbers***.
What are they?
How can you characterize them?

Write your answer
____
____
____
____
____

FIG. 6.3. Worksheet on Sieve of Eratosthenes.

sheets, students can explore concepts of divisibility, composite numbers, and prime decomposition, and expand their vocabulary.

This was one of the topics in the fourth-grade curriculum that Nurit had not yet taught. Moreover, it offered the potential for her to learn mathematics with her students. For example, Yolles (2001) reported that each year, when teaching the Sieve of Eratosthenes in fifth-grade class, she discovers new ideas. Extensive prior studies on teachers' content knowledge in mathematics with specific focus on number theory (e.g., Zazkis & Campbell, 1996; Zazkis & Liljedahl, 2004) have demonstrated the complexity of number theory for preservice elementary school mathematics teachers. In particular, Zazkis and Liljedahl (2004) investigated understanding of the concept of primality of numbers. They showed the complexity of the concepts especially at the stage of implementation in problem situations. They noted the Sieve of Eratosthenes as the way of "constructing primality by exclusion—by considering not what prime numbers are but what prime numbers are not. [This tool allows] 'sieving out' composite numbers from list of natural numbers in order to be left with primes only" (p. 182). Zazkis and

Liljedahl (2004) found that considering what primes are not is salient in teachers' understanding of prime numbers so the sieve has the potential to fit together with teachers' initial perceptions of the concept.

The Main Issues in the Interviews

During the prelesson meeting interviewer (Rachel[2]) asked Nurit to choose learning materials for her students and to explain her choice. Rachel invited Nurit to discuss the main concept introduced and practiced in the chosen materials, as well as planning the lessons in details. Nurit was asked about the concepts the students should know in advance in order to perform the tasks, and what new knowledge they should encounter during the lessons. As Nurit's mentor, Rachel felt free to ask questions in order to prompt Nurit's understanding, and Nurit took this conversation as quite natural. She also permitted videotaping the data, although asked that the camera be turned off when she was solving the tasks on the worksheets herself. Accordingly, the camera was turned off several times during the prelesson interview. When Nurit finished her solving, Rachel discussed the solutions with her and these discussions were videotaped.

Additionally to prelesson interviewing, Rachel observed two lessons in which Nurit taught the Sieve of Eratosthenes, wrote a detailed protocol, and took field notes. In the postinterview, Nurit was asked to analyze her planning versus teaching and to evaluate the interactive stage of the experiment. She was asked what her students had learned and what she herself had learned through teaching prime numbers and the Sieve of Eratosthenes.

LEARNING THROUGH TEACHING

This section describes the development of the teacher's knowledge of different kinds as evinced in the case of Nurit. The different accounts indicate the role that different forms of knowledge play in the development of teachers' knowledge through teaching (research question 2). The episodes in each account indicate the mechanisms of knowledge development (research question 1). Although this case presents a small piece of information collected in the bigger study, it exemplifies the way the data in the whole study were collected and analyzed, as well as partly illustrates the research findings as a whole.

---

[2]Rachel Dagan is an experienced elementary school mathematics teacher who is also working as a mentor. Nurit was one of her mentees. At the time of the experiment, Rachel was a graduate student who conducted this experiment as a part of her MA program within the framework of the ongoing study.

As shown earlier, Nurit chose to introduce the topic of prime numbers using the Sieve of Eratosthenes (see Fig. 6.4), so both the mathematical tool and the didactic approach to teaching prime numbers were unfamiliar to her. We observed her learning mathematics and pedagogy as she was planning the lesson through solving the problems that she was supposed to teach her students. Nurit needed a little guidance, with the help of which she solved most of the problems successfully.

Account 1: Refining and Expanding Systematic Knowledge: What Are Prime Numbers? How Does the Sieve Work?

As mentioned, Nurit was presented with a set of instructional materials for the topic of prime and composite numbers. Among other tasks she chose the worksheet with the Sieve of Eratosthenes to begin the topic. Nurit found some of the ideas completely new (e.g., the sieve and its use for the introduction of prime numbers) whereas she appeared to be acquainted with other ideas. Although prime numbers were not new ground to Nurit—she had learned the topic in her teacher training program and had taught it some years before, it still needed fine revision.

Throughout the prelesson interview, we (like Zazkis & Liljedahl, 2004) saw that the concept of primality of numbers was not easy for Nurit. She needed to better understand the concept in general and the structure of the sieve in particular. When asked whether she wanted to try to solve the worksheets, she said that they were easy, but when she

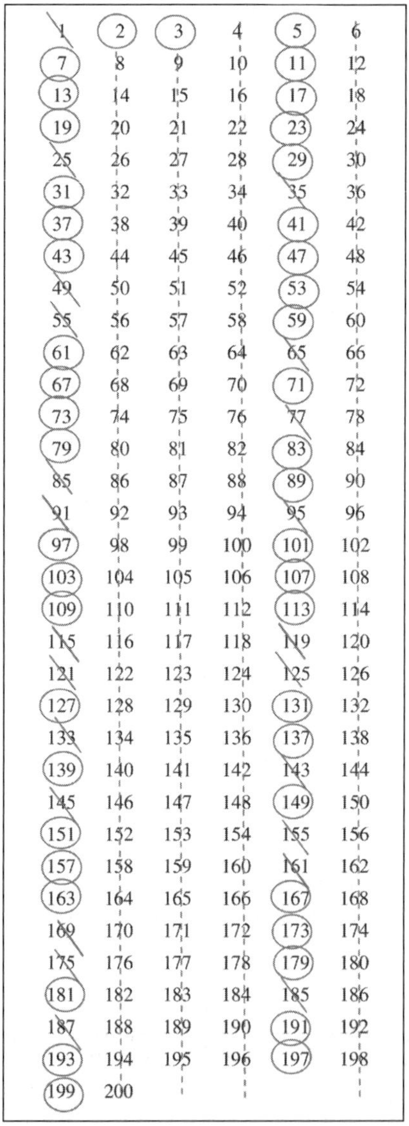

FIG. 6.4. Solution sheet: The Sieve of Eratosthenes.

started working on them, she asked us to turn off the camera.

Nurit a13: I don't need to learn it because *I know what prime numbers are*. But certainly *I have to recall all this*. I have not taught this topic for years so I have to recollect. [Asks to turn off the camera when she is solving the task.]

She worked on the tasks on "Sieve of Eratosthenes worksheet (see Fig. 6.4) for about 7 minutes, and then was ready to turn on the camera again. She was surprised to know "it was not so easy." She planned to distribute the worksheet among the students and to guide them toward the definition of prime numbers through this activity.

Episode 1.1: The Need to Be Prepared for the Lesson. The decision on how to start the topic was not easy. During the whole prelesson interview, Nurit seemed uncertain whether she should preface the work with a definition of prime numbers. Her need to know better than the students in order to teach them and to understand what she was going to teach the students was apparent. It stimulated her learning of mathematics and the development of her understanding of prime numbers in the course of the interview.

Right at the beginning of the planning meeting (prelesson interview) she said:

Nurit[3] a6: Obviously, before teaching a new topic I go through it myself, *to see that I remember* [what I knew] years ago, *that I know exactly*.

Later in the interview she returned to this issue several times:

Nurit a43, 46: I have not done this [the Sieve] ever.... *It will be good for me to learn a little as well*. So if somebody [a student] finds something it is worth knowing [in advance].

Nurit a65: I have to work with this carefully, to check these pages [worksheets] and *to solve them myself. I have to see that I understand everything* that we will do [during the lesson]. [I have to do this] because *if I don't understand I can't teach this as I should*.

The need to know better than students led Nurit toward learning the Sieve that she found unusual as well as toward refining the definition and the properties of the prime numbers that she considered familiar.

---

[3]Utterances from the prelesson interview are marked "ai." "i" signifies the number of Nurit's entrances into the conversation. Nurit's utterances in lesson 1 are marked "bi," those in lesson 2 are marked "ci," and those from the postlesson interview are marked "di."

Episode 1.2: Predicting Students' Difficulties. Nurit's need to learn the Sieve of Eratosthenes was rather natural because she "did not work with it beforehand." However, Nurit did not feel comfortable even telling herself that she needed to learn the topic. At some point in the interview, Nurit deflected the focus of the mathematical discussion to pedagogical issues. She kept clear of thinking about her own difficulties in understanding prime numbers and the Sieve of Eratosthenes. Instead, she concentrated on her uncertainty about the order of the stages of learning she was planning for her students.

Nurit a38: They [students] have to answer how many prime numbers there are among 200 natural numbers. So we say to them [the students], "These are prime numbers, now let's understand what prime numbers are."
Rachel: That's a possibility.
Nurit a39: Or I can do it opposite. I can explain these are prime numbers and then ask them, "How many prime numbers have you got here?"
Rachel: What kind of question you will ask?
Nurit a40–41: "Try to divide this number by other numbers" or "Try to see by which numbers these numbers can be divided?" [*Ironically*] *May be they will find something* .... I wonder if they could reach a conclusion on their own.

She was skeptical about the students' ability to perform the task and to define prime numbers based on the Sieve of Eratosthenes, as if projecting her own confusion onto how she expected her students' to perform in the lesson.

In common with many cases, Nurit talked about the students' ability to accomplish the tasks. She tried to predict their difficulties when approaching the tasks and how she could help them learn prime numbers. This attempt "to make students' life easier" led her to the conclusion that she needed to work more on the materials she was going to teach.

Nurit a67: They [the students] cannot understand what a prime number is without some explanation; they need help. *It will be very difficult for them* to understand what a prime number is.
Rachelle: So what do you think?
Nurit a69: I don't believe they'll ever get it on their own. I don't think so.
Rachelle: Why do you think they can't do it by themselves?
Nurit a70: It will be very difficult to them, without help ...
Nurit a75: You know, I didn't work enough with the worksheet.

It seems that Nurit strengthened her knowledge of prime numbers because contrary to her perception of the topic as difficult before the lesson, after it, she asserted that the material was easy and the students performed all the tasks without difficulty. For example, at the beginning of the postlesson interview Nurit said:

Nurit d5: They understood. It was easy for them. I gave them an exercise worksheet. You could see 90% of students in the class—if not more—understood the material. I don't know if they'll remember all this, but they understood .... We concluded, if these are prime numbers, what are composite numbers? *It was very easy for them ...*

Episode 1.3: Learning New Ideas from Students. Nurit's lesson was not rich in students' unexpected ideas, unlike some other lessons in the bigger study in which students solved a problem in an unexpected way, so that a teacher learned a new problem-solving strategy or even a new answer to the problem (the situation elaborated in Leikin & Dinur, 2003). One small episode from Nurit's lesson demonstrates the potential of learning from students' unexpected ideas:

Nurit b6: Consider numbers 3, 6, 9. What can you say about them?
Students: All of them are in the table of multiplication by 3.
This is a sequence.
This is an increasing sequence.
Nurit b6: Great! *Even I didn't think about this*. So these are the conclusions [you reached] from this small observation. You will need to write different conclusions in the worksheet you get.

Account 2: Starting With Mathematical Intuitions: Is the Power of a Prime a Prime?

Episode 2.1: Correcting Own Confusion. Even after performing the Sieve of Eratosthenes task, Nurit held a temporary intuition that a power of a prime is a prime. Rachel's request for providing an example of such a number let her understand that "she made a mistake" and she corrected it relying on Rachel's prompts.

Nurit a19: Well, *even I knew the topic it was a little [difficult]*. It is clear that 2 and 2 is one [writes 2:2 = 1], or 13 and thirteen is one. But I can get prime numbers when I multiply them by themselves several times, then they can be divided only by themselves. Right?

Rachel: *For example?*
Nurit a20: No they are not [prime numbers].
Rachel: What do you think?
Nurit a22: Because it is already multiplied by something, and then ... this means that only the number ...
Rachel: What happens to the number [when you multiply it by itself]?
Nurit a23: It becomes composite. So this means that I need only numbers that can be divided only by themselves.

Nurit's initially intuitive understanding that "the power of a prime number is prime" was corrected and transformed into formal knowledge that "the power of a prime is not prime." When the similar idea was raised by one of the students during the lesson, she had no difficulty identifying the students' confusion and helped them to resolve it.

Episode 2.2: Correcting Students' Mistake. When during the lesson one student's response was similar to the Nurit's own initial response: "By multiplying prime numbers several times we get primes," she noticed the students' confusion.

Student 3: What is a prime number?
Student 4: How do we know that a number is prime?
Nurit c11: We'll verify this in a minute. Seven, 13 [circles 7, 13].
Student 1: Twenty-five.

The need for scaffolding led her to continue the conversation as follows:

Nurit c13: [continues to circle numbers] 31, 37, 43. [Underlines 25 and 49].
Do you remember the divisibility criteria? Look at the columns and tell me what numbers are products of 3, 5 and 10? Take two minutes.
Student 2: What will I do with 2 minutes? I already know.
Nurit c14: Did you discover anything?
Student 2: Nothing is the product of nothing.
Student 4: Prime numbers are numbers that cannot be divided by anything
Student 2: You can cross out 49. It is divisible by 7.
Student 5: Prime numbers are divisible by themselves.
Nurit c15: Here you have it. I'll write and you copy it. [Writes:]
Definition: *A prime number is a number that may be divided by itself and by 1 only.*
Nurit c16: Let's do four accurately.

Student 6: It is divisible by itself
Nurit c17-18: [Writes 4:4 = 1]. Very good.
Student 2: By 2 also!!!!
Nurit: Here, it [4] is not alone. It also has 2 [as a divisor], so it is not prime! Do you understand?

As in many cases when students encountered a mathematical difficulty, Nurit, like other teachers in the larger study, had to find a fresh explanation. She linked prime and composite numbers and the divisibility criteria. This episode is a minor example of a teacher's mathematical idea that forms intuitively, and can be formalized further. As in other instances of mathematical scaffolding done by the teachers, Nurit developed a new mathematical connection, which gave rise to deeper understanding of the mathematical content being taught.

Note however that this connection is questionable: Zazkis and Campbell (1996) suggested that it may be harmful to introduce divisibility tests before the concept of divisibility has consolidated. They suggest postponing presentation of the divisibility criteria until conceptual understanding of divisibility is acquired.

Account 3: Missing an Opportunity: Space Left for Further Learning of Prime Numbers

Episode 3. 1: What Big Numbers Are Prime? It was evident during the prelesson interview, that Nurit did not know herself how to determine whether a big number is a prime.

Nurit a29: But still, what is still *problematic* here, is that when we take the example of 2 ... when I take small numbers students easily understand. But if I take 400 by 400 and we get 1 it will not look like a prime number [it is big].
Rachel: It is not
....
Nurit a32: Right. So we have to decide how to start [to make it easier].
Nurit a35: *It seems hard to me*. But I'll look at it [Sieve of Eratosthenes] once again.

Prime decomposition of numbers did not arise during either the prelesson interview or during the lessons that we observed. We tried to minimize our intervention in the teachers' learning-through-teaching process, so no prompts were given to Nurit by Rachel. Later on in the lesson, Nurit suggested that the students use the divisibility criterion as a tool for determining whether a number is composite (see Nurit c13 in Episode 2.2). Clearly, however, the criteria that the students learned (criteria for 2, 3, 5, and 10) could help only in relatively few cases.

Episode 3.2: What About 1? The first lesson was spent on small-group work with the worksheet and a whole-class discussion summarizing this work. Nurit concluded the first lesson by circling 2 and crossing out the columns of 2, 4, and 6 (see Fig. 6.4). In the second lesson, Nurit guided students toward defining prime numbers. She performed the task on the blackboard together with the students, because some of them did not have the worksheets with them.

Nurit c6: You were asked not to cross out 3 but the entire column [circles 3, crosses out other numbers in the column]
Student 1: *What about 1?*
Nurit c7-8: Very good. *We will cross out 1*. What is the conclusion? What can you say about prime and not prime numbers from the table?

At this point in the lesson Nurit did not ask the students why 1 is not a prime number. We expected her to discuss this issue with the students later, when they formulated the definition. But she did not return to this question.

Student 2: All even numbers are not prime numbers.
Nurit c9: What about 2? All even numbers but 2 are not prime numbers.
Nurit c10: Look at the table. Formulate definition.

Nurit ignored this student's question "*What about 1*?" and said, "*We cross out 1*" because the task called for crossing out the number (see Fig. 6.4), and, like other teachers in some cases, she did not notice the potential of students' questions for the continuation of the discussion. This was one of the missed opportunities for teacher's (as well as students') learning during a lesson.

Account 4: From Pedagogical Intuitions to Pedagogical Beliefs: "No New Topic at the End of the Year"

Nurit's intuitive reaction to our request to teach new material at the end of the school year turned into a belief. Because of her lack of experience in introducing the students to new concepts in mid-June (when the experiment took place), from the start of our experiment Nurit was anxious about it. She claimed "one cannot teach a new topic at the end of the year." She said this to Rachel in the pre and the postlesson interviews, as well as to the students:

Nurit b11: I know this is the end of the year and you're tired. But still, you have to read ...

Nurit's opening of the first lesson was rather problematic. Her instructions on dividing the class into small groups and on the mathematical task were rather vague. Organizing students into small groups took time; eventually they got to work in small groups, but their mathematical performance appeared very messy. One boy started filling in the numbers from the end—he put 200 in the bottom right-hand corner of the table. When the confusion became clear, his partners became angry with him because they had lost time on his account. Students in another group started to fill in the table from different directions, and "did not meet in the middle of the table." They claimed there was something wrong with the worksheet. Nurit simply put these complications down to the fact that it was the end of the school year and not a good time for teaching a new topic.

Although she was satisfied overall with the results of the lesson and said that most of the students understood the topic and liked the lesson (see Nurit d5 in the Episode 1.1), she failed to recognize this as "proof" of the possibility of teaching a new topic at the end of school year. She attributed a few flaws in the planning, unclear guidance to the students at the beginning of the first lesson, and organizational disorder—the students had to form small groups themselves—to the notion of "no new topic at the end of the year." This seems to have become a fixed belief in her mind:

Nurit d6: We did this in the whole-class discussion, not in small groups. That doesn't work at the end of the year.

Nurit d25: I knew it was not going to be easy for them at the end of the year. This topic is easy, not hard. But I know that each new topic has its difficulty.

Instead of countering her intuition "You cannot teach new material at the end of school year" with actual successful experience and critical analysis of the shortcomings of her instruction, she explained all the snags with one particular thought: Don't teach a new topic at the end of the year.

AWARENESS OF OWN LEARNING OF MATHEMATICS AND PEDAGOGY

Like most of the teachers who took part in this study, after the lesson Nurit was unaware of her own learning of prime numbers that had occurred during the teaching experiment. However, she clearly stated that she learned and enjoyed learning during the prelesson interview.

Nurit a103: I am always happy to get *new* materials, *to refresh what I have done in the past, it is also a pleasure for me to learn something new.*

She even reported to her students that she had learned what she was about to teach:

Nurit b1: You will do today something new and nice ... *I prepared the lesson together with Rachel, I learned it before the lesson.*

However, this awareness "dissolved" in her mind at the time of teaching. During the postinterview, she was aware that she had learned a lot of pedagogy; however, when she was asked if she felt she had learned mathematics, her response ignored the actual state of affairs:

Rachel: Do you feel that you learned [from this experiment]?
Nurit d30-31: I am aware that I haven't taught prime numbers for years. These materials [Eratosthenes' sieve] were unfamiliar to me. *It is nice teaching in a new way.* Usually we do [teach] prime numbers and composite numbers straightforwardly. I remember when I did it years ago, I did it quickly and less comprehensively. I invested more in this work [learning] and the students benefited more from the lessons ...
Next time I'll teach prime numbers with these worksheets. I think they are very nice.
Rachel: In other words, you mean the form [of teaching], not the mathematics?
Nurit d32: Because this mathematics is very simple. You can open everything [for new things], but ...

CONCLUDING REMARKS

This chapter presents one case taken from a multiple-case study that started on the premise that teachers learn when teaching: They develop different kinds of knowledge and become more proficient. The study was concerned in disclosing the mechanisms of this learning and the relationships between two dimensions of teachers' knowledge; its kind, and its forms. The methodology in this case study was identical to that employed in all the cases of the wider study. It included analysis of teaching mathematical topics that were neglected by the teacher or by the curriculum.

Nurit's case exemplifies the development of her knowledge while teaching prime numbers using the Sieve of Eratosthenes. The context for this case is the elementary school mathematics curriculum. It is especially interesting, because Nurit, like many other elementary teachers, was unaware of her learning mathematics while teaching. Even though the case of Nurit included several episodes in which she learned mathematics, she agreed that there are many opportunities for learning pedagogy when teaching and argued that elementary school mathematics "is simple" and does not have potential for her own learning. The choice of prime numbers in Nurit's case was based on two criteria: First, based on Rachel's communications with Nurit, we knew Nurit neglected this topic. Second, based on prior research (e.g., Zazkis & Campbell, 1996; Zazkis & Liljedahl, 2004), we believed that this topic had potential for the teacher's learning.

Derived from the observation that curricular changes take place very often in mathematics education, I assumed that teaching unfamiliar content is very natural for the teachers. Even when teaching familiar mathematical topics, teachers often meet neglected mathematical tasks. Use of teaching experiments of this kind was aimed at intensification of data collection.

One case cannot cover all possible teaching situations, but the case of Nurit partly answers the research questions of the whole study. At the same time, it illustrates nicely several ideas that emerged in the whole study, as well as demonstrates that teachers learn through teaching even elementary school mathematics. The chapter exemplifies several mechanisms for the development of teachers' knowledge in the course of teaching. It describes some roles that different forms of knowledge play in the development of teachers' knowledge through teaching.

How Teachers Develop Content (Mathematical) Knowledge in the Course of Teaching

When planning the lesson, Nurit clearly expressed her "need to know the material well enough" and her "need to predict students' possible difficulties, answers, and questions" (Episodes 1.1, 1.2). In other cases, similarly, these two interrelated themes were the main motives for teachers' learning during the planning stage. Nurit solved in advance problems that were to be solved by students during the lesson. Hence new pieces of information were collected and some familiar ideas were refined (account 1). The need to "know better than the students" stimulated Nurit's thinking about possible students' difficulties. When predicting them she reflected on her own uncertainties and thus answered her own questions when planning the lesson.

I argue that Nurit learned prime numbers while she prepared the lesson (without analyzing the depth and robustness of her developed understanding). The evidence for this is the promptness of her response to students' questions in the lesson on matters that in the prelesson interview that she needed time to resolve. Following the initial interview experience, and having prepared herself for the lesson by completing the worksheet and analyzing the obtained result, Nurit felt at ease with the material (Episode 2.2). This contrasted with what had been observed during the prelesson interview.

Observing students' mistakes during the lesson, Nurit directed her actions to scaffolding mathematical ideas. As in many other situations in the bigger study, Nurit searched for a new explanation to correct students' misunderstanding. She suggested that students use divisibility criteria to check whether a number was prime, so in the course of the lesson she constructed a new connection. She was surprised herself: She "thought about it only during the lesson." The other interesting source of teacher's learning during the lesson is learning from students' correct ideas unforeseen by the teacher. Unfortunately, Nurit's case did not include many episodes of learning from students' unpredicted correct ideas. However, one small episode (1.3) demonstrated that when interacting with students during the lessons, a teacher could learn new ideas presented by them. In several cases with other teachers, this kind of learning was significant (e.g., Leikin & Dinur, 2003).

One of the most important characteristics that determine learning through teaching is teachers noticing competence (Mason, 2002). When planning the lesson, Nurit noticed that the material was somewhat unfamiliar and thus felt a need to learn the material before the lesson. In a few other cases in the study, when teachers did not notice that the topic was unusual, they planned lessons without working carefully on the mathematical tasks, thus missing opportunities for their own learning. One example of such a missed opportunity is Episode 3.1. Although Nurit was curious about the question "what big numbers are prime?" she answered this question only partially and did not notice that there was still a potential to develop mathematical ideas related to this question. Noticing students' correct or incorrect answers during the lesson is no less important. Episode 3.2 is a situation in which Nurit did not notice the potential of the student's question and continued the discussion without addressing it.

Roles That Different Forms of Knowledge Play in the Development of Teachers' Knowledge Through Teaching

This research set forth two phenomena connected to the investigation of the roles that different forms of knowledge play in the development of teachers' knowledge through teaching: (1) Teachers' formal knowledge needs refining and enriching, and (2) teachers intuitive reactions may be transformed into formal knowledge or into beliefs.

The first phenomenon was rather natural for teaching unfamiliar (neglected) mathematical materials. As shown in account 1, when preparing the lesson Nurit clearly stated she needed to work on the materials, to refine the definition, and to understand how the Sieve works.

The second phenomenon was less obvious. As Atkinson and Claxton (2000) showed in "intuitive practitioner," many of the teachers' actions when teaching are intuitive and unplanned. With Nurit, her knowledge developed as the transformation of her intuitive reactions into formal knowledge or into beliefs (account 2). In terms of the relation between knowledge, intuitions, and beliefs suggested in the theoretical background to the chapter (Fig. 6.2), this research mainly outlined the transformation of mathematical intuitions into formal mathematical knowledge whereas pedagogical intuitions are transformed into beliefs. A mistaken intuition in the field of mathematics—"the power of a prime is prime"—was fortunately corrected by Nurit (Episode 2.1); thus her knowledge was refined. Nurit intuitively connected prime numbers and divisibility criteria during the lesson (Episode 2.2), so her own mathematical knowledge was enriched by a new mathematical connection.

Nurit's pedagogical intuitions seem to have exerted almost the opposite effect, as in many other cases observed in the larger study. She was asked to teach a new topic in June, but she had never previously introduced a new topic so late in the school year. Her intuitive response was, "You can't teach a new topic at the end of the year." Even though she liked the lesson and saw that the students understood the topic, she still did not accept this positive experience as evidence of the possibility of teaching a new topic at the end of school year. Instead, she took all the shortcomings of the lesson as support for her intuition, so that this opinion apparently became her belief. In several other cases in the larger study, a similar tendency was observed: Teachers tended to maintain that their pedagogical intuitions, which won no support during the lesson, were in fact correct, and their specific failure was purely by chance.

To conclude, this research demonstrates that teaching may serve as a learning environment for teachers themselves. Differently from other studies that stress the importance of reflection on action for the development of teachers' proficiency (e.g., Berliner, 1987) it shows that a teacher may learn mathematics and pedagogy when planning and performing a lesson. Naturally, teachers' learning through teaching is not absolute. There are different conditions that may either support or prevent such learning. Teacher education programs as well as mentoring should develop teachers' awareness of the possibility to learn from teaching and provide teachers with tools that may intensify teachers' learning through teaching.

ACKNOWLEDGMENT

I would like to thank Rachelle Dagan for her help in managing this teaching experiment and Nurit (pseudonym) for her agreement to participate in this research.

REFERENCES

Artzt, A. F., & Armour-Thomas, E. (2002). *Becoming a reflective mathematics teacher.* Mahwah, NJ: Lawrence Erlbaum Associates.

Atkinson, T., & Claxton, G. (Eds.). (2000). *The intuitive practitioner: On the value of not always knowing what one is doing.* Philadelphia, PA: Open University Press.

Berliner, D. C. (1987). Ways of thinking about students and classrooms by more and less experienced teachers. In J. Calderhead (Ed.), *Exploring teachers' thinking* (pp. 60–83). London: Cassell.

Chazan, D. (2000). *Beyond formulas in mathematics teaching: Dynamics of the high school algebra classroom.* New York: Teachers College.

Cobb, P., & McClain, K. (2001). An approach for supporting teachers' learning in social context. In F. L. Lin & T. J. Cooney (Eds.), *Making sense of mathematics teacher education* (pp. 207–231). Dordrecht, The Netherlands: Kluwer.

Cooney T. J. (2001). Considering the paradoxes, perils, and purposes of conceptualizing teacher development. In F. L. Lin & T. J. Cooney (Eds.), *Making sense of mathematics teacher education* (pp. 9–31). Dordrecht, The Netherlands: Kluwer.

Dewey, J. (1933). *How we think: A restatement of the relation of reflective thinking to the educative process.* Boston: D.C. Heath & Co.

Fischbein, E. (1982). Intuition and proof. *For the Learning of Mathematics, 3*(2), 9–18, 24.

Fischbein, E. (1984). The interactions between the formal, the algorithmic, and the intuitive components in a mathematical activity. In R. Biehler, R. W. Scholz, R. Sträßer, & B. Winkelmann (Eds.), *Didactics of mathematics as a scientific discipline* (pp. 231–245). Dordrecht, The Netherlands: Kluwer.

Franke, M. L., Carpenter, T. P., Levi, L., & Fennema, E. (2001). Capturing teachers' generative change: A follow-up study of professional development in mathematics. *American Educational Research Journal, 38,* 653–689.

Kennedy, M. M. (2002). Knowledge and teaching. *Teacher and Teaching: Theory and Practice, 8,* 355–370.

Lampert, M. (2001). *Teaching problems and the problems of teaching.* New Haven, CT: Yale University Press.

Lampert, M., & Ball, D. L. (1999). Aligning teacher education with contemporary K–12 reform visions. In L. Darling-Hammond & G. Sykes (Eds.), *Teaching as the learning profession: Handbook of policy and practice* (pp. 33–53). San Francisco: Jossey-Bass.

Leikin, R. (2004, January). *Learning through teaching: Intuitive and formal knowledge.* Paper presented at the Annual Meeting of Mathematics Educators in Israel: In Memoriam the late Prof. Efraim Fischbein. Tel Aviv University, Israel.

Leikin, R. (2005). Qualities of professional dialog: Connecting graduate research on teaching and the undergraduate teachers' program. *International Journal of Mathematical Education in Science and Technology, 36*(1-2), 237–256.

Leikin, R., Berman, A., & Zaslavsky, O. (2000). Learning through teaching: The case of symmetry. *Mathematics Education Research Journal, 12*(1), 17–36.

Leikin, R., & Dinur, S. (2003). Patterns of flexibility: Teachers' behavior in mathematical discussion. *Electronic Proceedings of the Third Conference of the European Society for Research in Mathematics Education.* http://www.dm.unipi.it/~didattica/CERME3/WG11

Leinhardt, G. (1993). On teaching. In R. Glaser (Ed.), *Advances in instructional psychology* (Vol. 4, pp. 1–54). Hillsdale, NJ: Lawrence Erlbaum Associates.

Lesh, R., & Kelly, A. E. (1994). Action-theoretic and phenomenological approaches to research in mathematics education: Studies on continually developing experts. In R. Biehler, R. W. Scholz, R. Sträßer, & B. Winkelmann (Eds.), *Didactics of mathematics as a scientific discipline* (pp. 277–286). Dordrecht, The Netherlands: Kluwer.

Ma, L. (1999). *Knowing and teaching elementary mathematics: Teacher's understanding of fundamental mathematics in China and the United States.* Mahwah, NJ: Lawrence Erlbaum Associates.

Mason, J. (1998). Enabling teachers to be real teachers: Necessary levels of awareness and structure of attention. *Journal of Mathematics Teacher Education, 1*(3), 243–267.

Mason, J. (2002). *Researching your own practice: The discipline of noticing*. New York: Falmer.

Roth, W. M. (1998). Science teaching as knowledgeability: A case study of knowing and learning during co-teaching. *Science Education, 82*(3), 357–377.

Scheffler, I. (1965). *Conditions of knowledge. An introduction to epistemology and education*. Glenview, IL: Scott, Foresman.

Schifter, D. (1998). Learning mathematics for teaching: From a teachers' seminar to the classroom. *Journal of Mathematics Teacher Education, 1*, 55–87.

Schön, D. A. (1983). *The reflective practitioner: How professionals think in action*. New York: Basic Books.

Shulman, L. S. (1986). Those who understand: Knowledge growth in teaching. *Educational Researcher, 15*(2), 4–14.

Simon, A. M. (1997). Developing new models of mathematics teaching: An imperative for research on mathematics teacher development. In E. Fennema & B. Scott-Nelson (Eds.), *Mathematics teachers in transition* (pp. 55–86). Mahwah, NJ: Lawrence Erlbaum Associates.

Skemp, R. R. (1987). *The psychology of learning mathematics.* Hillsdale, NJ: Lawrence Erlbaum Associates.

Steinbring, H. (1998). Elements of epistemological knowledge for mathematics teachers. *Journal of Mathematics Teacher Education, 1*(2), 157–189.

Stigler, J. W., & Hiebert, J. (1999). *The teaching gap: Best ideas from the world's teachers for improving education in the classroom.* New York: Free Press.

Strauss, A., & Corbin, J. (1990). Basics of qualitative research: Grounded theory procedures and techniques. Newbury Park, CA: Sage.

Wilson, S., Shulman, L., & Richert, A. E. (1987). "150" different ways of knowledge in teaching: Representations of knowledge in teaching. In J. Calderhead (Ed.), *Exploring teachers' thinking* (pp. 1–37). London: Cassell.

Yolles, A. (2001). Making connections with prime numbers. *Mathematics Teaching in the Middle School, 7*, 84.

Zazkis, R., & Campbell, C. (1996). Prime decomposition: Understanding uniqueness. *Journal of Mathematical Behavior, 15,* 207–218.

Zazkis, R., & Liljedahl, P. (2004). Understanding primes: The role of representation. *Journal for Research in Mathematics Education, 35,* 164–186.

Chapter 7

Learning Elementary Number Theory Through a Chain of Discovery: Preservice Teachers' Encounter with Pentominoes

Peter Liljedahl
Simon Fraser University

Elementary number theory concepts form the structural underpinnings of much of the arithmetic that is taught in the intermediate grades (Grades 4–7). Unfortunately, these underpinnings often remain an implicit part of a student's mathematical experiences, being displaced by a focus on the more obvious, and seemingly more important, algorithm for performing calculations. As such, students often graduate from their intermediate years with an impoverished view of, not only arithmetic but also mathematics; a view that promotes the memorization of facts and the mastery of algorithms over the comprehension of deeper mathematical structures (Borwein & Jörgenson, 2001). This is especially true for intermediate students' conceptions of division with remainder, the teaching of which is largely comprised of, if not completely displaced by, a focus on the algorithm for doing long division. As such, these students develop an impoverished view of division with remainder that hampers their ability to correctly answer even relatively simple story problems (Cai & Silver, 1995; Silver, 1992; Silver, Shapiro, & Deutsch, 1993). Ironically, this impoverished view is not restricted to the students. Research has shown that preservice elementary school teachers also have difficulty mastering concepts related to division with remainder (Campbell, 2002; Silver & Burkett, 1994; Zazkis, 1999, 2000; Zazkis &

Campbell, 1996a, 1996b). As such, any attempts to change the experiences of the students must start with changing the experiences of the teachers. In particular, teachers need to experience division with remainder concepts in a context other than in simple computation. They need to experience these concepts in a context that will allow for deeper exploration of the concepts themselves, and allow them to make explicit some of the understandings that are inherent in the algorithms they teach. Finally, teachers need to experience division with remainder concepts in an environment that is not only cognitively meaningful but also affectively rich, an environment that will foster positive beliefs and attitudes about mathematics in general (Liljedahl, 2002, 2004a, 2005) and division with remainder in particular.

In this chapter, I present a study that examines such an experience—a group of preservice teachers' exploring division with remainder in the context of a very specific type of problem-solving environment designed around, what I call, a *chain of discovery*.

A CHAIN OF DISCOVERY

Simply put, a chain of discovery is an experience in which a person, in the process of solving a mathematical problem, makes a series of mathematical discoveries. However, there are subtleties about such an experience that are lost through such a simple description. As such, a more detailed and descriptive explanation is necessary. In what follows, the idea of a chain of discovery is more thoroughly developed through the presentation of a progressively narrowing focus on specific types of mathematical problem solving. I begin with a cursory discussion of mathematical problem solving, and then examine three specific types of problem solving, of which chain of discovery is the third.

Mathematical Problem Solving

Much has been written about mathematical problem solving and to try to summarize it in a concise fashion is difficult; difficult primarily because opinions vary as to what constitutes a mathematical problem-solving experience, what constitutes mathematical problem solving, and even, what constitutes a mathematical problem. As such, rather than embark on either a synthesizing or a differentiating analysis of these varying understandings of mathematical problem solving, I present a brief working definition that will allow me to discuss the three subsequent specific cases of problem solving more effectively. First of all, mathematical problem solving can be thought of as being divided into two distinct, but related, processes—the logical and the extralogical; the logical processes will be dealt with in this section, the extra-logical processes will be treated in the next section.

The primary logical processes of problem solving are alternatively known as *problem solving by design* (Rusbult, 2000). The process begins with a clearly defined goal or objective after which there is a great reliance on relevant past experience, referred to as *repertoire* (Bruner, 1964; Schön, 1987), to produce possible options that will lead toward a solution of the problem (Poincaré, 1952). These options are then examined through a process of conscious evaluations (Dewey, 1933) to determine their suitability for advancing the problem toward the final goal. In very simple terms, problem solving by design is the process of deducing the solution from that which is already known.

Mayer (1982), Schoenfeld (1982), and Silver (1982) state that prior knowledge is a key element in the problem-solving process. Prior knowledge influences the problem solver's understanding of the problem as well as the choice of strategies that will be called on in trying to solve the problem. In fact, prior knowledge and prior experience is ALL that a solver has to draw on when first attacking a problem. As a result, all problem-solving heuristics must incorporate this resource of past experience and prior knowledge into their initial attack on a problem, and all do. Some heuristics refine these ideas, and some heuristics extend them. Of the heuristics that refine, none is more influential than the one created by George Pólya (1887–1985) in his book *How to Solve It* (1957), in which Pólya laid out a problem-solving heuristic that relies heavily on a repertoire of past experience. He summarized the four-step process of his heuristic as follows:

1. UNDERSTANDING THE PROBLEM
    - *First*. You have to understand the problem.
    - What is the unknown? What are the data? What is the condition?
    - Is it possible to satisfy the condition? Is the condition sufficient to determine the unknown? Or is it insufficient? Or redundant? Or contradictory?
    - Draw a figure. Introduce suitable notation.
    - Separate the various parts of the condition. Can you write them down?
2. DEVISING A PLAN
    - *Second*. Find the connection between the data and the unknown. You may be obliged to consider auxiliary problems if an immediate connection cannot be found. You should eventually obtain a plan of the solution.
    - Have you seen it before? Or have you seen the same problem in a slightly different form?
    - Do you know a related problem? Do you know a theorem that could be useful?
    - Look at the unknown! And try to think of a familiar problem having the same or a similar unknown.

- Here is a problem related to yours and solved before. Could you use it? Could you use its result? Could you use its method? Should you introduce some auxiliary element in order to make its use possible?
- Could you restate the problem? Could you restate it still differently? Go back to definitions.
- If you cannot solve the proposed problem, try to solve first some related problem. Could you imagine a more accessible related problem? A more general problem? A more special problem? An analogous problem? Could you solve a part of the problem? Keep only a part of the condition, drop the other part; how far is the unknown then determined, how can it vary? Could you derive something useful from the data? Could you think of other data appropriate to determine the unknown? Could you change the unknown or data, or both if necessary, so that the new unknown and the new data are nearer to each other?
- Did you use all the data? Did you use the whole condition? Have you taken into account all essential notions involved in the problem?

3. CARRYING OUT THE PLAN
   - *Third*. Carry out your plan.
   - Carrying out your plan of the solution, check each step. Can you see clearly that the step is correct? Can you prove that it is correct?
4. LOOKING BACK
   - *Fourth*. Examine the solution obtained.
   - Can you check the result? Can you check the argument?
   - Can you derive the solution differently? Can you see it at a glance?
   - Can you use the result, or the method, for some other problem?

The emphasis on auxiliary problems, related problems, and analogous problems that are, in themselves, also familiar problems is an explicit manifestation of this strategy. The use of familiar problems also requires an ability to deduce from these related problems a recognizable and relevant attribute that will transfer to the problem at hand. The mechanism that allows for this transfer of knowledge between analogous problems is known as analogical reasoning (English, 1997, 1998; Novick, 1988, 1990, 1995; Novick & Holyoak, 1991; Vosniadou & Ortony, 1989) and has been shown to be an effective, but not always accessible, problem-solving strategy.

Step four in Pólya's heuristic, looking back, is also a manifestation of utilizing prior knowledge to solve problems, albeit an implicit one. Looking back makes connections "in memory to previously acquired knowledge ... and further establishes knowledge in long-term memory that may be elaborated in later problem-solving encounters" (Silver, 1982, p. 20). That is, looking back is a forward-looking investment into future problem-solving encounters; it sets up connections that may later be needed.

Mathematical Discovery

Perkins (2000) made a clear distinction between problems that a person cannot solve and problems that a person has not yet solved. A problem that does not yield to a process of design is not necessarily unsolvable, but may merely be problematic. Such problems will require input from the extralogical processes in order for a solution to be found. The extralogical processes of problem solving are those processes that lie outside of the "theories of logical forms" (Dewey, 1938). Included in the cadre of the extralogical are such mysterious phenomena as intuition, imagination, insight, illumination, serendipity, and aesthetics, each of which may contribute to the solving of a problem in a fashion that may defy explanation, or more appropriately, defy logic. Also included in this collection are two phenomena that received much attention within the field of mathematics over the course of the last 100 years, creativity and discovery, both of which were born from the minds of French mathematicians.

In 1908, Henri Poincaré (1854–1912) gave a presentation to the French Psychological Society in Paris entitled "Mathematical Creation." This presentation, as well as the essay it spawned, stands to this day as one of the most insightful and thorough treatments of the topic of mathematical invention and put forth the proposition that the unconscious mind plays an invaluable role in the creative process. In particular, the anecdote of Poincaré's own discovery of Fuschian function transformations stands as the most famous contemporary account of mathematical creation.

> Just at this time, I left Caen, where I was living, to go on a geological excursion under the auspices of the School of Mines. The incident of the travel made me forget my mathematical work. Having reached Coutances, we entered an omnibus to go some place or other. At the moment when I put my foot on the step, the idea came to me, without anything in my former thoughts seeming to have paved the way for it, that the transformations I had used to define the Fuschian functions were identical with those of non-Euclidean geometry. I did not verify the idea; I should not have had the time, as, upon taking my seat in the omnibus, I went on with the conversation already commenced, but I felt a perfect certainty. On my return to Caen, for conscience' sake, I verified the results at my leisure. (Poincaré, 1952, p. 53)

Inspired by this presentation, Jacques Hadamard (1865–1963) began his own empirical investigation into mathematical invention. The results of this seminal work culminated in a series of lectures on mathematical invention at the École Libre des Hautes Etudes in New York City in 1943. These talks were subsequently published as *The Psychology of Mathematical Invention in the Mathematical Field* (Hadamard, 1945).

Hadamard's treatment of the subject of invention at the crossroads of mathematics and psychology was an entertaining, and sometimes humorous, look at the eccentric nature of mathematicians and their ritualistic practices. His work is an extensive exploration and extended argument for the existence of unconscious mental processes. To summarize, Hadamard took the ideas that Poincaré had posed and, borrowing a conceptual framework for the characterization of the creative process in general, turned them into a stage theory. This theory still stands as a viable and reasonable description of the process of mathematical discovery. In what follows, I present this theory, referenced not only to Hadamard and Poincaré, but also to some of the many researchers who's work has informed and verified different aspects of the theory.

The phenomenon of mathematical discovery, although marked by sudden illumination, consists of four separate stages stretched out over time, of which illumination is but one part. These stages are *initiation, incubation, illumination,* and *verification* (Hadamard, 1945). The first of these stages, the initiation phase, consists of deliberate and conscious work. This would constitute a person's voluntary, and seemingly fruitless, engagement with a problem and be characterized by an attempt to solve the problem by trolling through a repertoire of past experiences (Bruner, 1964; Schön, 1987). This is an important part of the inventive process because it creates the tension of unresolved effort that sets up the conditions necessary for the ensuing emotional release at the moment of illumination (Barnes, 2000; Davis & Hersh, 1980; Feynman, 1999; Hadamard, 1945; Poincaré, 1952; Rota, 1997).

Following the initiation stage, the solver, unable to come to a solution, stops working on the problem at a conscious level (Dewey, 1933) and begins to work on it at an unconscious level (Hadamard, 1945; Poincaré, 1952). This is referred to as the incubation stage of the inventive process and it is inextricably linked to the conscious and intentional effort that precedes it.

> There is another remark to be made about the conditions of this unconscious work: it is possible, and of a certainty it is only fruitful, if it is on the one hand preceded and on the other hand followed by a period of conscious work. These sudden inspirations never happen except after some days of voluntary effort which has appeared absolutely fruitless and whence nothing good seems to have come ... (Poincaré, 1952, p. 56)

After the period of incubation, a rapid coming to mind of a solution, referred to as illumination, may occur. This is accompanied by a feeling of certainty (Poincaré, 1952) and positive emotions (Barnes, 2000; Burton

1999; Rota, 1997). With regards to the phenomenon of illumination, it is clear that this phase is the manifestation of a bridging that occurs between the unconscious mind and the conscious mind (Poincaré, 1952), a coming to (conscious) mind of an idea or solution. However, what brings the idea forward to consciousness is unclear. There are theories on aesthetic qualities of the idea (Poincaré, 1952'; Sinclair, 2002), surprise/shock of recognition (Bruner, 1964), fluency of processing (Whittlesea & Williams, 2001), or breaking functional fixedness (Ashcraft, 1989) only one of which will be expanded upon here.

Poincaré proposed that ideas that were stimulated during initiation remained stimulated during incubation. However, freed from the constraints of conscious thought and deliberate calculation, these ideas would begin to come together in rapid and random unions so that "their mutual impacts may produce new combinations" (Poincaré, 1952, p. 61). These new combinations, or ideas, would then be evaluated for viability using an aesthetic sieve (Sinclair, 2002), which allowed through to the conscious mind only the "right combinations" (Poincaré, 1952, p. 62). It is important to note, however, that good or aesthetic does not necessarily mean correct. As such, correctness is evaluated during the fourth and final stage—*verification*.

Flow and Discovered Complexity

Flow (Csikszentmihalyi, 1996) is most simply described as the pleasurable state that a person may find himself or herself in when doing an activity. It is a state where one's actions are "automatic, effortless, yet it is also a highly focused state of consciousness" (p. 110), and enjoyment and engagement are at a maximum. Csikszentmihalyi identified nine key elements in people's descriptions of such states.

1. There are clear goals every step of the way.
2. There is immediate feedback on one's actions.
3. There is a balance between challenges and skills.
4. Attention is focused on one's actions.
5. Distractions are excluded from consciousness.
6. There is no worry of failure.
7. Self-consciousness disappears.
8. The sense of time becomes distorted.
9. The activity becomes satisfying in its own right.

Williams (2001) has taken Csikszentmihalyi's idea of flow and applied it to a specific instance of problem solving that she refers to as *discovered complexity*. Discovered complexity is a state that occurs when a problem solver, or a group of problem solvers, encounter complexities that were not evident at the onset of the task and that are within their zone of proximal development (Vygotsky, 1978). This occurs when the solver(s) "spontaneously formulate a question (intellectual challenge) that is resolved as they work with unfamiliar mathematical ideas" (p. 378). Such an encounter will capture and hold the engagement of the problem solver(s) in a way that satisfies the conditions of flow. What Williams' framework describes is the deep engagement that is sometimes observed in students working on a problem-solving task during a single-problem solving session. What it does not describe, however, is a student's willingness to return to the same task, again and again, over an extended period of time (several days or weeks) until the problem is solved. Such willingness requires a different theoretical framework to explain, a framework built from a chain of discovery.

A Chain of Discovery

Willingness to engage in a problem-solving activity resides within a student's affective domain. It may reside within the student's beliefs and attitudes or it may reside within the student's emotions (McLeod, 1992). Beliefs are just that, what students believe; what they believe to be true about mathematics and what they believe about their ability to do mathematics. Beliefs about mathematics are often based on their own experiences with mathematics. For example, beliefs that mathematics is "difficult," "useless," "all about one answer," or "all about memorizing formulas" stem from experiences that have first introduced these ideas and then reinforced them.

A qualitatively different form of belief is with regards to a person's beliefs in their ability to do mathematics, often referred to as efficacy, or self-efficacy. Self-efficacy, like the aforementioned belief structures, is a product of an individual's experiences with mathematics. Self-efficacy with regard to mathematics has most often been dealt with in the context of negative belief structures (Ponte, Matos, Guimarães, Cunha Leal, & Canavarro, 1992) such as "I can't do math," "I don't have a mathematical mind," or even "girls aren't good at math."

Attitudes can be defined as "a disposition to respond favourably or unfavourably to an object, person, institution, or event" (Ajzen, 1988, p. 4) and can be thought of as the responses that students have to their belief structures. That is, attitudes are the manifestations of beliefs. For example, beliefs such as "math is difficult," "math is useless," or "I can't do math" may result in an attitude such as "math sucks." A belief that "math is all about formulas" may manifest itself as an attitude of disregard for explanations in anticipation of the eventual presentation of a formula.

Emotions, on the other hand, are relatively unstable (Eynde, De Corte, & Verschaffel, 2001). They are rooted more in the immediacy of a situation or a task and as a result, are often fleeting. Students with generally negative beliefs and attitudes can experience moments of positive emotions about a task at hand or, conversely, students with generally positive outlooks can experience negative emotions.

So, the willingness of a student to engage with a problem during a single problem-solving session may, in fact, be due to a temporarily heightened positive emotional response to the situation. This is certainly in keeping with the foundation of "enjoyment" existing within the frameworks of both Csikszentmihalyi (1996) and Williams (2001). To extend this engagement across several sessions, however, beliefs and attitudes about the task must also be in a positive state. Within some student this may already be the case in that the student may have positive beliefs and attitudes about either problem solving in general, or the topic in which the problem is set in particular. Within other students, however, this may be more problematic.

For a student who has already developed negative beliefs and attitudes about mathematics and/or problem solving, engagement in a problem-solving task across many sessions will be difficult. Research has shown that, although negative beliefs and attitudes are slow to form in a learner, they are equally slow to change once formed (Eynde, De Corte, & Verschaffel, 2001). Ironically, change is most often achieved through the emotional dimension in that repeated positive experiences will eventually produce positive beliefs and attitudes. However, change is slow and will generally require a large number of successive successes before any long-term change is observed. As such, the "enjoyment" that may be experienced in the first session of an extended series of problem solving will likely not effect enough change to encourage engagement in subsequent sessions.

However, there is one mechanism by which change in attitudes and beliefs can be quickly realized. Research has shown that the experience of discovery in the context of mathematical problem solving has an immediate and powerfully transformative effect on learners beliefs about mathematics and their ability to do mathematics and the attitudes that govern their behavior in the context of doing mathematics (Liljedahl, 2002, 2004a, 2005). These changes can be as wide sweeping as "I now like mathematics," but are more likely to fall into the domain of "I can do this." The exact mechanism within discovery that facilities this change is not clear but it may be linked to feelings of certainty (Burton, 1999; Fischbein, 1987), bliss (Rota, 1997), and the release of the tension from unresolved effort that occurs at the flash of illumination; "that moment when the connection is made, in that synaptic spasm of completion when the thought drives through the red fuse, is our keenest pleasure" (Harris, 2000, p. 132). Extrapolating this positive effect across a series of discoveries will not only magnify change in the affective domain, but will also maintain the engagement of the student in the problem-solving task across several distinct sessions.

A chain of discoveries will facilitate such an extrapolation. It occurs when a problem solver, or a group of problem solvers, encounter successive discoveries in the course of solving a problem over an extended period of time requiring an extended number of problem-solving sessions. Each new discovery provides the solver(s) with new information and new tools to aid in the advancement toward an eventual solution, as well as provides the necessary changes in the affective domain to sustain engagement across these many sessions.

OCCASIONING A CHAIN OF DISCOVERY

The idea that discovery in general, and a chain of discovery in particular, can be orchestrated has to be qualified with the fact that such experiences are largely dependent on chance; intrinsic and extrinsic chance. Intrinsic chance deals with the luck of coming up with an answer, of having the right combination of ideas join within your mind to produce a new understanding. This was discussed by Hadamard (1945) as well as by a host of others under the name of "the chance hypothesis." Extrinsic chance, on the other hand, deals with the luck associated with a chance reading of an article, a chance encounter with an individual or some piece of mathematical knowledge, any of which contributes to the eventual resolution of the problem on which one is working. However, the idea that mathematical discovery often relies on the fleeting and unpredictable occurrence of chance encounters is starkly contradictory to the image projected by mathematics as a field reliant on logic and deductive reasoning. Ironically, this contributory role of chance emerged from a study done with prominent research mathematicians in which a portion of Hadamard's (1945) famous survey was resurrected in order to solicit responses pertaining to illumination, creativity, and insight (Liljedahl, 2004a, 2004b). The upshot of this strong dependence on chance means that the orchestration of discovery can best be described as the *occasioning* (Kieren, Simmt, & Mgembelo, 1997) of discovery. That is, the environment for such an experience can be orchestrated, but the experience itself cannot.

In general, an environment conducive to discovery is one that provides interaction, time, and a rich task. Research has shown that much more is achieved in the context of problem solving from interacting with others, whether that interaction is collaborative, or merely conversational (Liljedahl, 2004a). People learn from each other and they learn in conjunction with others. In fact, research mathematicians claim that they are much more likely to garner something that will help them solve a particular problem through a conversation with someone than from reading something (Liljedahl, 2004a).

Time is also an important element in the discovery process. A discovery is much more likely to occur after much time has been dedicated to the solving of a problem, and is much more likely to occur at the moment when a problem is revisited (Liljedahl, 2004a). This means that, to begin with, a lot of time has to be provided in the form of allocated time as well as extended deadlines. It also means that within this allocation of time, there needs to be many opportunities to revisit the problem.

Finally, a discovery is much more likely to occur in the context of working on a good problem. Schoenfeld (1982) has come up with a list of characteristics of a good problem, summarized as follows:

- The problem needs to be accessible. That is, it is easily understood, and does not require specific knowledge to get into.
- The problem can be approached from a number of different ways.
- The problem should serve as an introduction to important mathematical ideas.
- The problem should serve as a starting point for rich mathematical exploration and lead to more good problems.

However, a good problem for the purposes of occasioning a discovery is slightly different than what is more generally considered a good problem. Perkins (2000) felt that what is important is that a problem be "problematic." That is, the problem must cause the solver to get stuck. For the purposes of occasioning discovery and engaging the participants in the process of problem solving, the first two of Schoenfeld's points as well as Perkins' requirement have the greatest relevance. Research has also shown that problems that produce early success and problems that stimulate conversation are the most likely to foster a discovery (Liljedahl, 2004a). For the purposes of occasioning a chain of discovery in particular, the last of Schoenfeld's conditions is most relevant.

CHAIN OF DISCOVERY, ELEMENTARY NUMBER THEORY, AND PRESERVICE TEACHERS: A RESEARCH STUDY

In this section, I present a study that exemplifies a chain of discovery, as well as demonstrates the depth to which mathematical content can be accessed through such a chain. This study details the results of a group of preservice elementary school teachers working to solve a problem steeped in one particular elementary number theory concept, the concept of division with remainder. Cai and Silver (1995), Silver et al. (1993), and Silver (1992) examined this concept as it related to "real life" situations. Campbell (2002) looked at the concept as it related to partitive and quotitive division and the division algorithm. Both showed that participants' understandings of remainder were firmly rooted in the algorithm for performing long division and any understanding of the remainder in terms of the nature of the dividend was rare and limited. My research contributes in exemplifying how a concrete understanding of the remainder as it relates to the nature of the dividend can be achieved or extended through the chain of discovery experienced in attempting to solve the pentominoe problem.

The Participants

Participants in this study are preservice elementary school teachers enrolled in six different offerings of a Designs for Learning Elementary Mathematics course for which I was the instructor, spanning six semesters and 4 years and involving a total of more than 180 students. The course, which runs for 13 weeks, is specifically designed to provide preservice teachers with the foundational knowledge and skills requisite to teach mathematics at the elementary (k–7) level. As such, the focus of the course is more on teaching than on the specifics of mathematics content, although such content is often used as the context in which teaching and learning are discussed. Because of the nature of the courses, and the timing with which they are offered, the majority of the students who take these courses were enrolled in the teacher preparation program offered at the university.

Each offering of the course enrolls between 28 and 35 students with about 90% being female. In addition, the vast majority of the students in each offering are extremely fearful of having to take mathematics and even more so of having to teach mathematics. This fear resides, most often, within their negative beliefs and attitudes about their ability to learn and do mathematics. At the same time, however, as apprehensive and fearful of mathematics as these students are, they are extremely open to, and appreciative of, any ideas that may help them to become better mathematics teachers. They are willing to engage in discussions regarding pedagogy and philosophy and are willing to reflect on how to apply these in their teaching.

The Problem

The problem,[1] as given to the participants, can be seen in Fig. 7.1.

Clearly this is a problem that places much focus on the elementary number theory concepts of divisibility and division with remainder.

The Context

Although the participants are eventually provided with a hard copy of the problem (as presented earlier) their actual exposure to the problem is done in pieces. To begin with, they are simply given the task of determining how

[1]This problem was presented to me by Ralph Mason. For many other problems related to pentominoes, see Solomon Golomb's (1996) *Polyominoes*.

A pentominoe is a shape that is created by the joining of five squares such that every square touches at least one other square along a full edge. Now consider a 100s chart.

1	2	3	4	5	6	7	8	9	10
11	12	13	14	15	16	17	18	19	20
21	22	23	24	25	26	27	28	29	30
31	32	33	34	35	36	37	38	39	40
41	42	43	44	45	46	47	48	49	50
51	52	53	54	55	56	57	58	59	60
61	62	63	64	65	66	67	68	69	70
71	72	73	74	75	76	77	78	79	80
81	82	83	84	85	86	87	88	89	90
91	92	93	94	95	96	97	98	99	100

If a pentominoe is placed somewhere on a 100's chart will the sum of the numbers it covers be divisible by 5? If not, what will the remainder be? Explain how you can know **quickly**!

FIG. 7.1. The pentominoe assignment.

many pentominoes there are. This is done as an in-class, group problem-solving activity in which they use small square tiles as manipulatives. Once the class as a whole settles on the fact that there are 12 such shapes (see Fig. 7.2), they are presented with the problem as shown in Fig. 7.1.

I allow them to discuss the problem for a few minutes within their groups and then I answer any clarifying questions they may have. Invariably, one of the questions that is asked has to do with what is meant by "quickly." Rather than answer this question, I perform a demonstration. I ask a student to come up and, using their square tile manipulatives, construct a pentominoe of their choice anywhere on a 100s chart that is set up on the overhead projector. While they do this, I close my eyes. On their command, I look at what they have constructed and immediately (within 1 or 2 secs) tell them what the remainder is. This is an important aspect of the problem for it is the participants' attempts to perform the calculation "quickly" that invariably moves them through a chain of discoveries; at each link learning something more about the problem.

The participants then work in groups, both in and out of class, over the remainder of the course to solve the problem. I also make the announce-

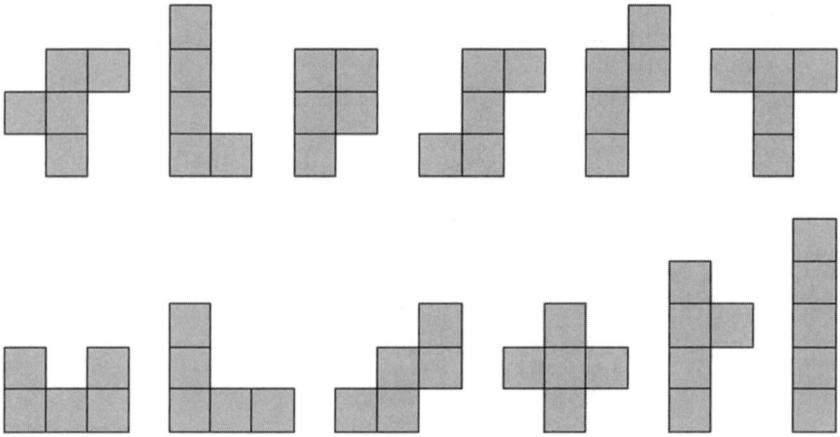

FIG. 7.2. The 12 pentominoes.

ment that they are not likely to arrive at the final solution until they have spent more than 10 hours working on it. This announcement, like the requirement of quickly helps them to move through the chain of discoveries instead of getting caught at one link with the thought that they are done. It also helps to prepare them for the work that they will be required to put into the problem. Finally, the 10-hour minimum serves as a challenge to try to solve the problem in less time.

The Source of the Data

Data for this study comes from a variety of different journals that each participant kept during their enrollment in the course. These journals ranged from personal problem-solving journals, to group problem-solving journals, to personal reflective journals. Personal problem-solving journals were used to track the participants' own trials, progress, successes, failures, and breakthroughs detailing both the mathematics that they are grappling with and the feelings they are experiencing as they work to solve the pentominoe problem. The group problem-solving journals were very similar in nature to the aforementioned personal journals except that they were to be used to track the efforts of the group as they worked on the problem. As such, entries in these journals detailed the goings-on of the group meetings and the collaborative work that occurred in these meetings. Finally, the personal reflective journals provided a place in which they respond to prompts regarding their thoughts and experiences during the course. The entries in this journal are much more contemplative in nature and provide insights into how the participants thinking has changed through their problem-solving experiences within the course.

Together these three journals provide an invaluable picture of what went on while the participants struggled to solve the pentominoe problem, both on their own and in groups. All the journals were collected and sorted according to the groups that worked together. The relevant entries from all of these journals were then coded for emergent themes and cross-checked with journals from other group members.

Results and Analysis

This analysis produced a wealth of data on the phenomenon of chain of discovery in the context of solving the pentominoe problem. In what follows I present, with the use of exemplifying excerpts, this chain of discoveries as it emerges from the data. I illustrate the relevant division with remainder concepts that the participants encounter at each link within the chain as well as discuss the nature of the discovery that predicate the advancement to each link.

Position Doesn't Matter. The first, and most obvious discovery that the students make is that the position of the pentominoe on the 100s chart does not affect the remainder. This very simple discovery seems to serve three very distinct and useful purposes. First, as already mentioned, in a problem as complex and as lengthy as this one, it is useful to have some early success to motivate the students to keep working (Liljedahl, 2004a). In this case, once the remainder for each of the 12 pentominoes has been determined for each of their four orientations, the students have, in essence, established a solution. Given the inelegance of this method, they realize that it is not *the* solution, but having at least *a* solution makes them feel somewhat safe. Rebecca states this nicely in her journal:

> Right away we figured out that the position doesn't matter. This was a huge relief to us because the problem suddenly went from something that seemed completely undoable to something that we could mange.

More important, however, the realization that the position is irrelevant gets the students focusing on the remainder. Although this is an explicit requirement of the problem, the students do not immediately focus on this one attribute alone. This is likely due to the fact that their first attempts at this problem are very mechanical; placing a pentominoe on the 100s chart, summing the numbers that are covered, and dividing by 5. This series of steps produces two results—a quotient and a remainder. Moving a given pentominoe around changes many quantities within this procedure; it changes the numbers that are covered, it changes their sum (the dividend), and it changes the quotient. It does not, however, change the remainder. Until this discovery is made, the students tend to track all of the seemingly relevant information (the specific numbers to be summed, the sum of these numbers, the quotient, and the remainder) as they move their pentominoes around the 100s chart. Once the discovery is made, however, they dispense with tracking all but the remainder. This letting go of the particulars is a positive step toward occasioning discovery (Liljedahl, 2004a, 2004b, 2005).

Finally, the student's attempts to explain why the position does not matter leads to some very useful understanding with regards to division with remainder. In this particular case, moving a pentominoe vertically by one cell on the 100s chart changes the dividend by 50, and moving it horizontally by one cell changes the dividend by 5. When the divisor is 5, neither of these changes in the dividend will produce a change in the remainder. This property is nicely demonstrated in Dave's comments:

> We realized right away that the remainder was the same no matter where we put the shape. It took a lot longer to figure out why this was so. It wasn't until we got a hint from Peter [the instructor] that we started thinking about moving the shape just slightly. If we take a shape, say the L,[2] and we put it on the table so that it covers the 25, 35, 45, 46, and 47 then the sum will be 198. If we now move the L one square to the right then the numbers that are covered will be 26, 36, 46, 47, and 48. We just increased each of the numbers by 1. So, the new total is now 203, 5 more than it was before. But this doesn't matter, because when we divide by 5 we still have a remainder of 3. Likewise, if we move the L down then the numbers that are now covered will be 35, 45, 55, 56, and 57. Basically, we increased each of the numbers by 10, for a total of 50. Adding 50 to a number doesn't change the remainder because 5 goes into 50.

This is a specific case of a more general property of division with remainder; namely that changing the dividend by multiple of the divisor does not change the remainder.[3]

Stemming from the realization that the position of the pentominoe does not affect the remainder emerges a strategy of intentionally placing the pentominoes in specific regions of the grid in order to simplify the calculations. In particular, students begin to place the pentominoes where the in-

---

[2]The participants usually named the pentominoe shapes according to the letter or symbol that they most closely resembled. The L, for example, is for this participant the pentominoe .

[3]Notice that Dave presents all of his arguments in arithmetic terms (with specific numbers and operations) as opposed to algebraic terms. Of the more than 180 participants who have worked on this problem, none have submitted solutions that use algebraic symbolization.

dividual numbers are the smallest; that is, in the top left hand corner of the 100s chart. In fact, adopting this strategy allows them to reduce the 100s chart to a 5 × 5 grid (Fig. 7.3a). This repositioning is often accompanied with a strategy of ignoring the 10s digit of the individual cell numbers in the chart as seen in the comments of Jennifer:

> We can just ignore the 10s part of the number because adding on all those 10s doesn't change the last number, and it is the last number that determines the remainder.

Sometimes this also results in an explicit renumbering of the chart using only the one's digits (see Fig. 7.3b).

Combining these two strategies produces a strategy of focusing on only the top left-hand corner of the renumbered grid (Fig. 7.3c) as invoked by Jonathan.

1	**2**	**3**	**4**	**5**	6	7	8	9	10
11	**12**	**13**	**14**	**15**	16	17	18	19	20
21	**22**	**23**	**24**	**25**	26	27	28	29	30
31	**32**	**33**	**34**	**35**	36	37	38	39	40
41	**42**	**43**	**44**	**45**	46	47	48	49	50
51	52	53	54	55	56	57	58	59	60
61	62	63	64	65	66	67	68	69	70
71	72	73	74	75	76	77	78	79	80
81	82	83	84	85	86	87	88	89	90
91	92	93	94	95	96	97	98	99	100

FIG. 7.3a. Reducing the grid.

1	2	3	4	5	6	7	8	9	0
1	2	3	4	5	6	7	8	9	0
1	2	3	4	5	6	7	8	9	0
1	2	3	4	5	6	7	8	9	0
1	2	3	4	5	6	7	8	9	0
1	2	3	4	5	6	7	8	9	0
1	2	3	4	5	6	7	8	9	0
1	2	3	4	5	6	7	8	9	0
1	2	3	4	5	6	7	8	9	0
1	2	3	4	5	6	7	8	9	0

FIG. 7.3b. Renumbering the grid.

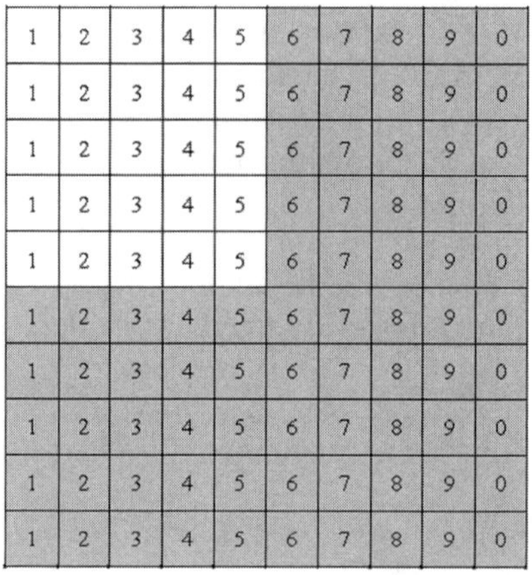

FIG. 7.3c. Reducing and renumbering the grid.

> Because the position of the pentominoe doesn't matter we'll just put it in the top left-hand corner of our new chart. This way the numbers are small and every possible pentominoe will still fit in there.

These methods of streamlining the calculations of the remainders allow the students to dispense with the use of a calculator. In all of these cases, however, these strategies all involve, in one form or the other, the explicit summing of the dividend as a means to access the remainder. Although a much improved method for calculating the remainder, further improvement is needed if the students want to be able to "very quickly" perform the calculation. This improvement comes when the students begin to consider how each piece of the pentominoe will contribute to the final remainder, as opposed to the dividend, and is discussed in great detail in a subsequent section.

Orientation Does Matter[4] Whereas the position of the pentominoe does not affect the remainder, the orientation of the pentominoe does. This discovery is often made simultaneously with the previous one, and although more difficult to explain than the irrelevance of position, does lead to some interesting patterns regarding symmetry, reflections, and rotations.

The participants very quickly discover that those shapes that are symmetrical about a vertical axis (cf. ) have a remainder of zero, and those shapes that are equally weighted or balanced on both sides of a verti-

[4] In strictly mathematical terms, "orientation" is part of "position." However, as a theme emerging from the data, the orientation of the pentominoe is distinct from its position.

cal axis (cf. ) also have a remainder of zero. Both of these discoveries are nicely demonstrated in the words of Veronica:

> When we looked at the pentominoes that had no remainder, the first thing we noticed was that from the midpoint of the shape there was always an equal number of blocks on each side. Sometimes the shapes were symmetrical like the +, the U and the T. Other times the shapes were not symmetrical like the Z. We decided that all of these shapes were somehow balanced and the blocks must cancel each other out and make a remainder of zero.

A further discovery is that for those shapes that do not have a remainder of zero, the reflection of the pentominoe about any vertical axis will produce a remainder that is the 5s compliment of the original remainder. For example, while 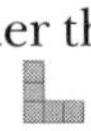 has a remainder of 3, 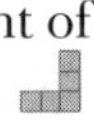 has a remainder of 2. Betty makes this observation using a different set of pentominoes than the one provided in the previous example.

> When we looked at the little L we noticed that its remainder and the remainder of its reflection added up to 5 (Fig. 7.4). When we looked at more shapes we discovered that mirror reflections of pentominoe remainders *always* add up to 5.

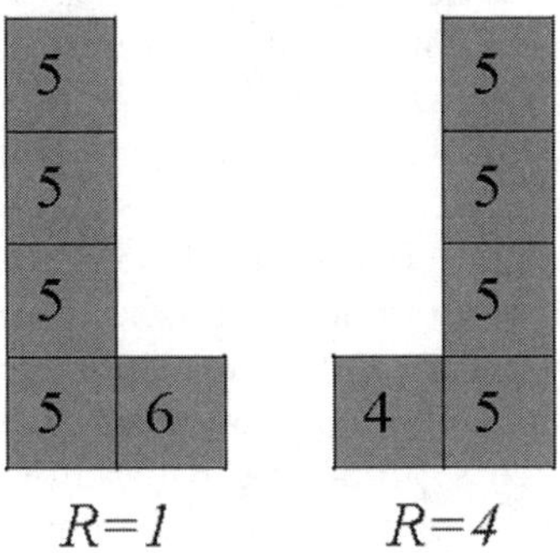

FIG. 7.4. Betty's example of reflection complements.

Further to this point of reflections, it is also quickly discovered that any reflections across a horizontal axis produce no change in the remainder.

Similar to the complements produced by reflection are the complements produced by half-turn rotations. That is, the half-turn rotation of a given pentominoe will also produce a 5s compliment of the original remainder. So, while 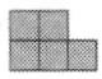 has a remainder of 4, 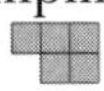 has a remainder of one. John demonstrates this realization in his problem-solving journal in the context of exploring rotations in general:

> We looked at each of the 18 pentominoe shapes[5] with their 4 rotations. We started to notice that if a shape had no remainder then neither did its 180° rotation. Then we noticed that if the remainder was not 0, then the 180° rotation would be the opposite of the original remainder. For example, the P has a remainder of 2 and its 180° rotation has a remainder of 3. They always add up to 5.

Although not evident in the previous passage, this "noticing" of the pattern was much more of a discovery than John indicates here. In his reflective journal, John writes about this discovery:

> We had all been working on the pentominoe problem separately. We had each taken three of the pentominoe shapes and had calculated what each of the remainders were depending on how you turned them. When we met that first time we shared our results and it was just a mess. It was pretty obvious that the symmetrical shapes had a remainder of zero. Reva also had noticed that the reflections of her shapes always added up to five and when we checked our own shapes we found this was always true. We were beginning to talk about how we would divide up the rest of the work when all of a sudden it hit me—Reva's idea about reflection also worked for rotation. It was right there on the paper and it just jumped out at me all of a sudden ...

These four discoveries (symmetry, balance, reflection complements, and rotation complements) emerge through the students' attempts to catalogue the remainders for each pentominoe and its various orientations. The identification of these four properties serves to reduce the solution space from many seemingly unrelated remainders to less than many sometimes related remainders. However, more connections and patterns are needed in order to be able to decode the remainder of a given pentominoe placed in a given position, and a given orientation somewhere on the 100s chart. Unlike the discovery concerning position, however, insights into these four discoveries are not forthcoming at this point in the students' problem-solving efforts. That is, although they are able to identify these patterns, they are not yet able to explain them. Furthermore, they are often not able to explain these patterns until a dynamic understanding of the problem is achieved near the very end of the problem-solving process (see following section).

Summing of Remainder. An altogether different strategy for calculating the remainder comes from considering how each cell would individually contribute to the remainder. That is, when division with remainder is performed on a sum, then the remainder of the sum will equal the sum of

[5]When the original 12 shapes are considered in conjunction with their reflections, then there are 18 distinct shapes.

the remainders.[6] This is another property that this problem seems to make explicit for the students, as can be seen in the following reference from a group journal:

> It was our fourth meeting and so far, we hadn't gotten very far. When Jessica arrived, she showed us something that she had been trying on her own. She had renumbered the entire grid 1, 2, 3, 4, 0, 1, 2, 3, 4, 0. She explained that these numbers are the remainders that each number on the grid has. This was a much easier grid to work with.... Why it works—when we take five numbers and add them up and divide by five to get the remainder, we are really just looking at how many groups of five we can remove from the total and how much is left over. But, this means that we have to first add a bunch of groups of five first and then remove them later. Renumbering the grid, all we really do is remove the groups of five first. So, a number like 23 is really just 4 groups of 5 and a 3 left over. If our pentominoe shape covers the 23 we only have to worry about the 3.

A closer look at Jessica's reflective journal reveals how she came up with this idea.

> At our last meeting we had decided that we were going to renumber the 100s chart by ignoring the first part of the number. This was fine, I was doing this in my head anyway. I had made a new chart on my computer, and I was filling in the rows [columns] one by one. When I got to the fifth row [column] I was suddenly struck by the idea that I didn't need this row. It was going to be full of 5s anyway and 5s didn't change the remainder. This was a very strange idea. It felt both right and wrong at the same time ...

The students who chose to follow this strategy all renumbered the 100s chart such that each cell is replaced with its remainder in division by five (see Fig. 7.5a).

At first glance this may not seem to produce much of a change in the calculation of the remainder, especially if the student is invoking a strategy of placing the pentominoes in the top left-hand corner of the grid as in the previous section. Consider, for example, the pentominoe 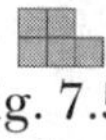 being placed in the top left-hand corner of the renumbered grid in Fig. 7.5b and in the top left-hand corner of Fig. 7.3c. In both cases, the calculation of the remainder would involve the summing of (1, 2, 1, 2, 3) and the eventual division of the sum by 5 and the extraction of the remainder. So, at the level of arithmetic, the renumbering of the grid presented in Fig. 7.5a produces no change. However, at the level of conception, this renumbering of the grid is profound. As will be seen in subsequent sections, this renumbering is a necessary step toward the liberation of the solution from the particulars of the

[6]That is, if $x \equiv a \pmod{k}$, and $y \equiv b \pmod{k}$, then $x + y \equiv a + b \pmod{k}$.

1	2	3	4	0	1	2	3	4	0
1	2	3	4	0	1	2	3	4	0
1	2	3	4	0	1	2	3	4	0
1	2	3	4	0	1	2	3	4	0
1	2	3	4	0	1	2	3	4	0
1	2	3	4	0	1	2	3	4	0
1	2	3	4	0	1	2	3	4	0
1	2	3	4	0	1	2	3	4	0
1	2	3	4	0	1	2	3	4	0
1	2	3	4	0	1	2	3	4	0

FIG. 7.5a. Reducing the grid to remainders.

numbers. More immediately, however, it allows the students to place the pentominoes in a more strategic position on the grid (see Fig. 7.5c); a placement that, yet again, reduces the number of calculations that are necessary. This is seen in Leslie's work:

> We realized that we can really speed up our calculations if we put the pentominoe in the centre of the 100s chart. This way some of the squares that are covered would be zeros.

In general, the initial urge for the students is to place the pentominoes on the "remainder grid" so that the left most column of the pentominoe sits on the "zero line." However, they quickly come to the realization that certain columns of the pentominoe are better placed on the "zero line." Leanne nicely articulates this:

1	2	3	4	0	1	2	3	4	0
1	2	3	4	0	1	2	3	4	0
1	2	3	4	0	1	2	3	4	0
1	2	3	4	0	1	2	3	4	0
1	2	3	4	0	1	2	3	4	0
1	2	3	4	0	1	2	3	4	0
1	2	3	4	0	1	2	3	4	0
1	2	3	4	0	1	2	3	4	0
1	2	3	4	0	1	2	3	4	0
1	2	3	4	0	1	2	3	4	0

FIG. 7.5b. Placing pentominoe on new grid.

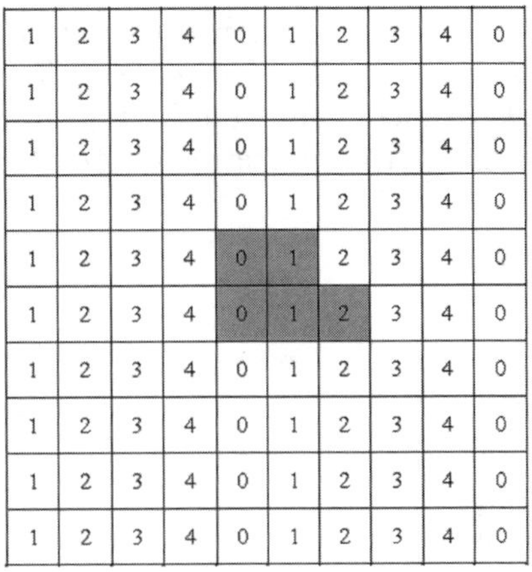

FIG. 7.5c. Replacing pentominoe on new grid.

At first we were just putting the pentominoes on the fives column and to the right. This way we started with zero and then just counted over how many 1s we had, how many 2s we had, etc. Then Alyssa suddenly suggested that for some shapes we should put the pentominoe down so that we have the most number of squares on the fives column. For example, look at the backwards L (Fig. 7.6). The calculation is easier if we put the long part on the zeros.

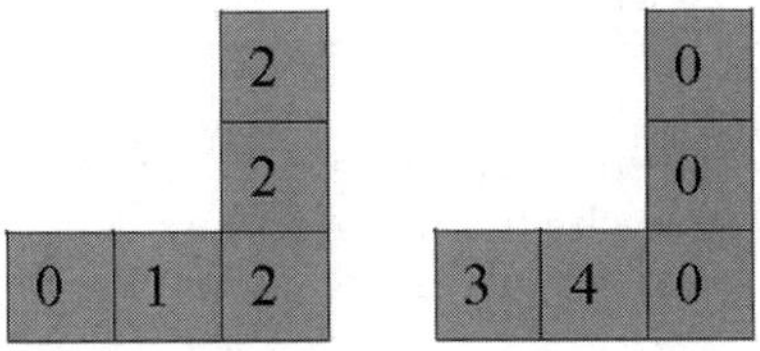

FIG. 7.6. Leanne's example of the backwards "L."

Looking at Alyssa's reflective journal reveals more about the nature of this idea:

> My most powerful discovery occurred when we were working on the pentominoe problem in class. We had just figured out that we could renumber the grid 1, 2, 3, 4, 0, 1, 2, 3, 4, 0. At first I had argued against this. It didn't seem to be that different from what we were doing and it was just one more thing that we had to explain to Peter [the instructor]. Anyway, we had decided that we would try using this grid because it kept the numbers small. We were putting the different shapes on the zero line and then racing to see how quickly we could figure out what the remainder was. Someone put the L shape down and suddenly I saw that were doing it wrong. We didn't have to start on the zero line. We could start anywhere, and the best thing would be if the zero line knocked out the most number of squares ...

Negative Remainders. The next discovery comes in the realization that a remainder of 4 is, in fact, the same as a number being one short of a multiple of 5, or a remainder of –1. Likewise a remainder of 3 is the same as a remainder of –2. This produces yet another renumbering of the grid as seen in Fig. 7.7.

		-2	-1	0	1	2			
		-2	-1	0	1	2			
		-2	-1	0	1	2			
		-2	-1	0	1	2			
		-2	-1	0	1	2			
		-2	-1	0	1	2			
		-2	-1	0	1	2			
		-2	-1	0	1	2			
		-2	-1	0	1	2			
		-2	-1	0	1	2			

FIG. 7.7. Renumbering of the grid with negative remainders.

For the students, this renumbering speeds up the calculation of remainders even more because it allows them to "cancel out" numbers as Joanne explains:

> So, now I can place the pentominoes so that some of the squares cancel each other out. For example, I can place 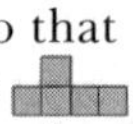 with the longest column on the zero (Fig. 7.8). This way the 1 and the –1 cancel out leaving just the 2.

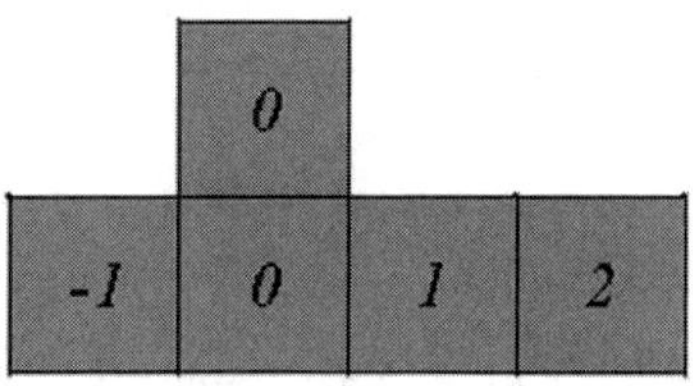

FIG. 7.8. Joanne's example of "canceling out."

Looking to Diane, one of Joanne's group mates, shows where this idea came from:

> We were doing the adding up the remainders thing when I put the backwards L down on the grid. Everyone right away said that the remainder was 4, but I wanted to say it was one short. Then I had my AHA! Remainder of 4 is the same as remainder –1 and remainder 3 was the same as remainder –2. I knew right away that this was right, and I knew it was important. I explained this to my partners and we decided that we should try to renumber the grid *again*. It felt great. I was finally contributing to the group.

It also allows them to very easily explain their earlier observations regarding symmetry, balance, reflection complements, and rotation complements presented in an earlier section. This is seen in Angela's comments:

> We noticed in the remainder zero group that there are always the same number of squares on the left as on the right. We now see that these cancel each other out because the ones on the right have a positive remainder and the ones on the left have a negative remainder.

Dynamic Summing of Remainder. The final, and perhaps ultimate discovery for the students occurs when they liberate themselves from the grid altogether and develop what I have come to call a "dynamic perception" of the pentominoes. Up until this stage, the students have worked with the pentominoes as static objects and they have manipulated everything else. They have played with the pentominoes position and orientation, and they have played with the contents of the 100s chart. In this final stage of the chain of discoveries, the students begin to manipulate the pentominoes themselves. This is well demonstrated in Colleen's words:

> Gary suggested that we start with the I, which we know has a remainder of zero. Then we give each step away from our starting position a value of 1 or −1 depending on which direction it moves. So, to get the remainder for the stairway we move one block to the left (−1) and two blocks to the right (+2). So, the remainder will be +1 (Fig. 7.9).

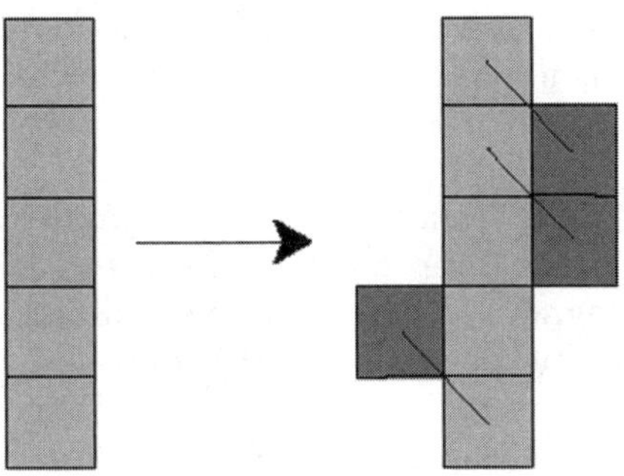

FIG. 7.9. Colleen's example of dynamic summing.

The connection of this very nonnumber-based explanation to the concept of division with remainder comes from the fact that every horizontal move of a single cell will change the dividend by only one, thereby changing the remainder by only one (positive or negative). At the same time, every vertical move of a block will change the dividend by a total of 10 and hence have no effect on the remainder. So, in essence the remainder can be calculated by simply tracking the horizontal moves.

This reshaping of the "I" into the various pentominoe shapes is an explanation for something that occurs in a much more innate manner. The constant exposure to the various shapes and the prolonged effort to solve the problem leads the participants to spontaneously leave the workspace of the grid behind. An example of this can be seen in Sharon's journal:

> Laura suggested that we use the blocks to work on the pentominoes. She started making the shapes of the pentominoes with the cubes and she would move one block to make different pentominoe. As Laura moved the blocks Jen and I started calling out the remainders. At first we were just calling them out of memory, yet as we did this and Laura moved blocks we suddenly started to see a pattern or a system.

This discovery, like the previous discovery of moving the axis allows the students to very quickly explain the properties they discovered in the early parts of their exploration. This is nicely demonstrated by Elain's explanation:

> We are always counting how many moves we make to the left and to the right. So, if we make the same number of moves to the left as to right then the remainder will be zero. This explains why all the symmetric shapes (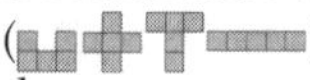) and all the balanced shapes () have no remainder.

DISCUSSION

The chain of discovery presented earlier can be summarized in the diagram presented in Fig. 7.10. It should be noted that although many of the participants traverse this chain exactly as presented here, not all do. Some of the participants bypass the discoveries represented in cells 2 and 4 as shown with the arrows. It should also be noted, however, that all participants make the discoveries summarized in cells 1, 3, and 5, and furthermore, they make the discoveries in this order.

The initial realization that position doesn't matter is more of an immediate observation than a discovery and hence is represented differently than any of the other discoveries in the diagram. Nonetheless, this realization leads to some number theory-related concepts regarding the contribution of the 10s part of a number to the remainder.

The first real discovery that is made pertains to the patterns that are observed with respect to the remainders of reflected and rotated pentominoes. Although this discovery remains unexplained until much later in the solution process, the property of complements is a result of a relative consideration of contribution to the remainder. That is, although all numbers contribute to the dividend, some numbers increase the remainder and some numbers decrease the remainder.

The next two discoveries that are made have to do with the realization that it is advantageous to consider the contribution that each number makes to the remainder instead of the contribution that each number

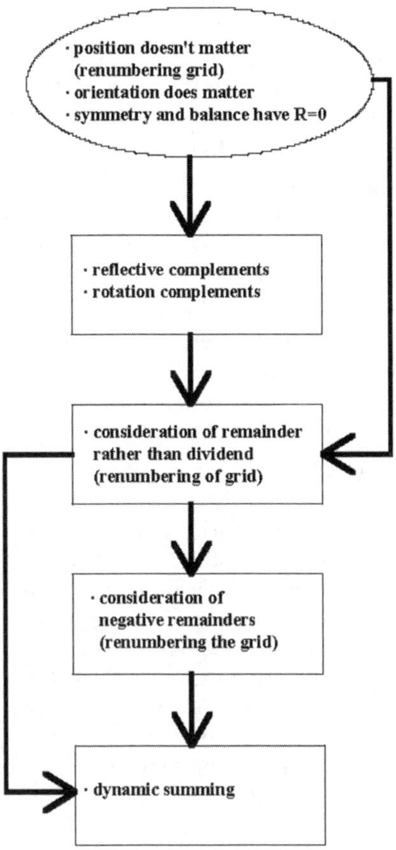

FIG. 7.10. Summary of the pantominoe chain of discovery.

makes to the dividend. This comes from the fact that the remainder of a sum will be the same as the sum of the remainders. The initial discovery brings to light this idea in the context of only positive contributions that a number can make to the remainder and is represented in the third box of Fig. 7.10. The secondary discovery of this concept comes when it is realized that contributions can be both positive and negative in nature and is represented in the fourth box of the diagram.

Finally, the last box represents the discovery that all discussions of remainders can be liberated from the 100s chart all together. This then leads to the dynamic construction of pentominoe shapes and tracking of the effect that each move of a block has on the final remainder.

The chain of discoveries detailed in the previous sections and summarized here is more than just a set of understandings that are achieved. Each new understanding is punctuated with a discovery. John sees the pattern that exists between rotated pentominoe shapes "jump out" at him when he suddenly saw how the property of reflection also applied to rotation. Jessica was "suddenly struck" by the realization that individual numbers contributed to the eventual remainder and thus warranted the renumbering of the grid to reflect these contributions. Alyssa "suddenly saw that [her group] was doing it wrong" and that the pentominoes could be placed more strategically for the calculation of remainders. Dianne had an AHA! and "knew right away" that remainders could be seen as being both positive and negative. Finally, Sharon "suddenly started to see a pattern" that the dynamic construction of pentominoes was producing. For the participants, the concepts that are presented to them at each link in the chain of discovery arrive in a flash of illumination on the heels of much deliberate effort and periods of incubation. These moments of discovery also bring with them all of the positive emotions that normally accompany such experiences and serve to motivate the participants to keep going. This state of constant motivation is nicely captured in Lisa's reflective journal:

> Of all the problems that we worked on my favourite was definitely the pentominoe problem. We worked so hard on it, and it took forever to get the final answer. But I never felt like giving up, I always had confidence that we would get through it. Every time we got stuck we would just keep at it and suddenly one of us would make a discovery and we would be off to the races again. That's how it was the whole time—get stuck, work hard, make a discovery—over and over again. It was great. I actually began to look forward to our group sessions working on the problem. I have never felt this way about mathematics before—NEVER! I now feel like this is ok, I'm ok, I'll BE ok. I can do mathematics, and I definitely want my students to feel this way when I teach mathematics ...

This journal entry also highlights the overall enjoyment that the participants felt regarding the problem as a whole. In fact, in surveying the various classes of students who were exposed to this problem between 75% and 90% of the students identified this problem as their favorite. Furthermore, between 85% and 95% of the students identified it as the most memorable problem they were exposed to, and 100% of the students identified it as a problem within which they experienced moments of discovery.

CONCLUSION

Research has shown that preservice teachers' understanding of division with remainder is rooted in their procedural understanding of the long division algorithm (Campbell, 2002; Zazkis, 1999, 2000; Zazkis & Campbell, 1996a, 1996b). In particular, preservice teachers are often unable to determine the remainder without explicitly performing long division, even in cases where the remainder can be discerned from the nature of the dividend. In the study presented in this chapter, similar difficulties also presented themselves. However, these difficulties were overcome as the participants moved through a chain of discoveries occasioned by their work on the pentominoe problem. As they moved through the chain of discoveries, their understanding of remainder evolved, beginning with an understanding of remainder as one of the results of long division and ending with the understanding of how changes to the dividend contribute to changes in the remainder.

However, a chain of discovery is much more than just a cognitive experience. It is also an affective experience capable of changing even very robust negative beliefs and attitudes within a population of apprehensive and mathematically timid preservice elementary school teachers. The particular chain of discovery presented in this chapter is no different. Not only was the cognitive domain of the participants changed with respect to division with remainder but so too was the affective domain. The pentominoe problem engaged the participants, and it sustained that engagement through the changes in the attitudes and beliefs that it fostered.

REFERENCES

Ajzen, I. (1988). *Attitudes, personality, and behaviour*. Milton Keynes, England: Open University Press.

Ashcraft, M. (1989). *Human memory and cognition*. Glenview, IL: Scott, Foresman.

Barnes, M. (2000). 'Magical' moments in mathematics: Insights into the process of coming to know. *For the Learning of Mathematics, 20*(1), 33–43.

Borwein, P., & Jörgenson, L. (2001, February). Visible structures in number theory. *MAA Monthly*, 897–910.

Bruner, J. (1964). *Bruner on knowing*. Cambridge, MA: Harvard University Press.

Burton, L. (1999). The practices of mathematicians: What do they tell us about coming to know mathematics? *Educational Studies in Mathematics, 37*(2), 121–143.

Cai, J., & Silver, E. (1995). Solution processes and interpretations of solutions in solving a division-with-remainder story problem: Do Chinese and U.S. students have similar difficulties? *Journal for Research in Mathematics Education, 26*(5), 491–97.

Campbell, S. R. (2002). Coming to terms with division: Preservice teachers' understanding. In S. R. Campbell & R. Zazkis (Eds.), *Learning and teaching number theory: Research in cognition and research* (pp. 15–40). Westport, CT: Ablex.

Csikszentmihalyi, C. (1996). *Creativity: Flow and the psychology of discovery and invention*. New York: HarperCollins.

Davis, P., & Hersh, R. (1980). *The mathematical experience*. Boston, MA: Birkhauser.

Dewey, J. (1933). *How we think*. Boston, MA: Heath.

Dewey, J. (1938). *Logic: The theory of inquiry*. New York: Holt.

English, L. (1997). Children's reasoning processes in classifying and solving comparison word problems. In L. D. English (Ed.), *Mathematical reasoning: Analogies, metaphors, and images* (pp. 191–220). Mahwah, NJ: Lawrence Erlbaum Associates.

English, L. (1998). Reasoning by analogy in solving comparison problems. *Mathematical Cognition, 4*(2), 125–146.

Eynde, P., De Corte, E., & Verschaffel, L. (2001). Problem solving in the mathematics classroom: A socio-constructivist account of the role of students' emotions. *Proceedings of 25th Annual Conference for the Psychology of Mathematics Education* (Vol. 4, pp. 25–32). Utrecht, Netherlands: Utrecht University.

Feynman, R. (1999). *The pleasure of finding things out.* Cambridge, MA: Perseus Publishing.

Fischbein, E. (1987). *Intuition in science and mathematics: An educational approach.* Dordrecht, Netherlands: Kluwer.

Golomb, S. (1996). *Polyominoes.* Princeton, NJ: Princeton University Press.

Hadamard, J. (1945). *The psychology of invention in the mathematical field.* New York: Dover.

Harris, T. (2000). *Hannibal.* New York: Dell.

Kieren, T., Simmt, E., & Mgembelo, J. (1997). Occasioning understanding: Understanding occasioning. In J. Dossey, J. Swafford, M. Parmantie, & A. Dossey (Eds.), *Proceedings of the 26th International Conference for Psychology of Mathematics Education—North American Chapter* (vol. 2, pp. 627–634). Bloomington- Normal, IL.

Liljedahl, P. (2002). The 'AHA moment': Students insights in learning mathematics. *Proceedings of the 24th International Conference for Psychology of Mathematics Education—North American Chapter* (Vol. 2, pp. 887–896). Atlanta, GA.

Liljedahl, P. (2004a). *The AHA! experience: Mathematical contexts, pedagogical implications.* Unpublished doctoral dissertation, Simon Fraser University, Burnaby, British Columbia, Canada.

Liljedahl, P. (2004b). Mathematical discovery: Hadamard resurrected. *Proceedings of the 28th International Conference for Psychology of Mathematics Education* (Vol. 3, pp. 249–256). Bergen, Norway.

Liljedahl, P. (2004c). Improving problem-solving journals: The mathematician, the narrator, and the participant. *Proceedings of the 26th International Conference for Psychology of Mathematics Education—North American Chapter* (Vol. 2, pp. 439–446). Toronto, Ontario.

Liljedahl, P. (2005). Mathematical discovery and affect: The effect of AHA! experiences on undergraduate mathematics students. *International Journal of Mathematical Education in Science and Technology, 36*(2–3), 219–236.

Mayer, R. (1982). The psychology of mathematical problem solving. In F. K. Lester & J. Garofalo (Eds.), *Mathematical problem solving: Issues in research* (pp. 1–13). Philadelphia, PA: Franklin Institute Press.

McLeod, D. (1992). Research on the affect in mathematics education: A reconceptualization. In D. A. Grouws (Ed.), *Handbook of research on mathematics teaching and learning* (pp. 575–596). New York: Macmillan.

Novick, L. (1988). Analogical transfer, problem similarity, and expertise. *Journal of Educational Psychology: Learning, Memory, and Cognition, 14*(3), 510–520.

Novick, L. (1990). Representational transfer in problem solving. *Psychological Science, 1*(2), 128–132.

Novick, L. (1995). Some determinants of successful analogical transfer in the solution of algebra word problems. *Thinking and Reasoning, 1*(1), 5–30.

Novick, L., & Holyoak, K. (1991). Mathematical problem solving by analogy. *Journal of Experimental Psychology, 17*(3), 398–415.

Perkins, D. (2000). *Archimedes' bathtub: The art of breakthrough thinking*. New York: Norton.

Pólya, G. (1957). *How to solve it* (2nd ed.). Princeton, NJ: Princeton University Press.

Poincaré, H. (1952). *Science and method*. New York: Dover.

Ponte, J., Matos, J., Guimarães, H., Cunha Leal, L., & Canavarro, A. (1992). Students' views and attitudes towards mathematics teaching and learning: A case study of curriculum experience. *Proceedings of 16th Annual Conference for the Psychology of Mathematics Education* (Vol. 2, pp. 218–225). Durham, NH: University of New Hampshire.

Rota, G. (1997). *Indiscrete thoughts*. Boston, MA: Birkhauser.

Rusbult, C. (2000). *An introduction to design*. Retrieved December 3, 2003, from http://www.asa3.org/ASA/education/think/intro.htm#process

Schoenfeld, A. (1982). Some thoughts on problem-solving research and mathematics education. In F. K. Lester & J. Garofalo (Eds.), *Mathematical problem solving: Issues in research* (pp. 27–37). Philadelphia, PA: Franklin Institute Press.

Schön, D. (1987). *Educating the reflective practitioner*. San Francisco, CA: Jossey-Bass.

Silver, E. (1982). Knowledge organization and mathematical problem solving. In F. K. Lester & J. Garofalo (Eds.), *Mathematical problem solving: Issues in research* (pp. 15–25). Philadelphia, PA: Franklin Institute Press.

Silver, E. (1992). Referential mappings and the solution of division story problems involving remainders. *Focus on Learning Problems in Mathematics, 14*(3), 29–39.

Silver, E., & Burkett, M. (1994, April). *The posing of division problems by preservice elementary school teachers: Conceptual knowledge and contextual connections*. Paper presented at the 1994 Annual Meeting of the American Educational Research Association. New Orleans, LA.

Silver, E., Shapiro, L., & Deutsch, A. (1993). Sense making and the solution of division problems involving remainders: An examination of middle school students' solution processes and their interpretations of solutions. *Journal for Research in Mathematics Education, 24*(2), 117–135.

Sinclair, N. (2002). The kissing triangles: The aesthetics of mathematical discovery. *International Journal of Computers for Mathematics Learning, 7*(1), 45–63.

Vosniadou, S., & Ortony, A. (1989). Similarity and analogical reasoning: A synthesis. In S. Vosniadou & A. Ortony (Eds.), *Similarity and analogical reasoning* (pp. 1–15). New York: Cambridge University Press.

Vygotsky, L. (1978). *Mind in society: The development of higher psychological processes*. Cambridge, MA: Harvard University Press.

Whittlesea, B., & Williams, L. (2001). The discrepancy-attribution hypothesis: The heuristic basis of feelings of familiarity. *Journal for Experimental Psychology: Learning, Memory, and Cognition, 27*(1), 3–13.

Williams, G. (2001). Sustained overlap in zones of proximal development? If so, how did that happen? *Proceedings of the 25th International Conference for Psychology of Mathematics Education* (Vol. 1, p. 378). Utrecht, Netherlands.

Zazkis, R. (1999). Intuitive rules in number theory: Example of "the more of A, the more of B" rule implementation. *Educational Studies in Mathematics, 40*(2), 197–209.

Zazkis, R. (2000). Factors, divisors and multiples: Exploring the web of students' connections. *Research in Collegiate Mathematics Education, 4,* 210–238.

Zazkis, R., & Campbell, S. R. (1996a). Divisibility and multiplicative structure of natural numbers: Preservice teachers' understanding. *Journal for Research in Mathematics Education, 27*(5), 540–563.

Zazkis, R., & Campbell, S. R. (1996b). Prime decomposition: Understanding uniqueness. *Journal of Mathematical Behavior, 15*(2), 207–218.

Chapter 8

The Number-Theoretic Experience of 12- to 15-Year-Olds in a Calculator Environment: The Intertwining Coemergence of Technique and Theory

Carolyn Kieran
Université du Québec à Montréal

José Guzmán
Centro de Investigación y de Estudios Avanzados del Instituto Politécnico Nacional

The new instruments of mathematical work are of interest ... because they permit students to develop new techniques that constitute a bridge between tasks and theories.

—Lagrange (2000, p. 17)

Research involving 12- to 15-year-old students is often oriented toward a better understanding of their development of algebraic, geometric, and proportional reasoning. The emergence of number-theoretic ideas among students of this age is rarely studied. If the high school mathematics programs of the province of Quebec (MEQ, 1993) and of Mexico (SEP, 2000) are at all indicative of the content of programs elsewhere, then it should be the case that students in, for example, their first year of high school (12 to

13 years old) are developing a conceptual understanding of factors, multiples, prime numbers, and divisibility. But the attention given to such topics in existing textbooks and other professional resources would seem to be very limited. Even less exists on the ways in which technology might be harnessed by mathematics teachers in encouraging the growth of these concepts in their students. Thus, it is not only researchers who have paid little heed to the evolving of number-theoretic ideas among middle and early secondary school students; so too have curriculum developers. The result is that we have scant knowledge regarding number-theoretic thinking among students of this age and how it might be fostered. The research that is described in this chapter is aimed at opening up a small part of this understudied area in mathematics education. It focuses on the emergence of some basic number-theoretic ideas related, in particular, to factors, multiples, and divisibility, among 12- to 15-year-olds, and on the role that can be played by calculator tools in the intertwining development of students' theoretical and technical knowledge in this domain.

THEORETICAL FRAMEWORK

Recent technology-based research (e.g., Artigue, 2002; Guin & Trouche, 1999; Lagrange, 2000) related to the coemergence of theoretical and technical knowledge has its roots in the anthropological theory of Chevallard (1999) and the ergonomic theory of Vérillon and Rabardel (1995). According to Vérillon and Rabardel, whose work draws upon Vygotskian theory, tools exist as both artifacts (physical objects) and instruments (psychological constructions). These two researchers suggest that, when students appropriate tools for themselves, and thereby transform artifacts into instruments, they are not simply learning tool techniques in response to given tasks. They are actually developing conceptually while they are perfecting their techniques with the tool.

Jean-Baptiste Lagrange (2000) has argued that the dialectical interaction occurring in this process involves three components: technique, task, and theorizing. These three ts are, in fact, based on an earlier elaboration by Chevallard (1999) that comprised four components—technology being the fourth t, but the sense in which Chevallard used the term is quite different from the way in which we do so.

Chevallard (1999) defined technique as "a way of accomplishing, of carrying out tasks" and pointed out that a technique "is not necessarily algorithmic or quasi algorithmic" (p. 225). In his anthropological theory, he separates technique from the discourse that justifies/explains/produces it, which he refers to as *technology*.[1] But he also admits that this type of dis-

[1]Our use of *Technology*, in contrast, refers to computers and calculator tools.

course is often integrated into technique, and has pointed out that such technique can be characterized in terms of theoretical progress. This is a key aspect of our research. In other words, the notion of a relation existing between technique and its underlying theoretical rationale is an important one that forms the backbone of our perspective on the coemergence of theory and technique. Lest our opening quotation from Lagrange (2000) suggest a static image of technique spanning the two shores of task and *existing* theory, let us hasten to clarify that the main point is rather the supporting role played by technique in the bootstrapping development of theory by the student.

Lagrange (2002) has expressed his perspective on the interrelationship of technique, task, and theory as follows:

> Within this dynamic, tasks are first of all problems. Techniques become elaborated relative to tasks, then become hierarchically differentiated. Official techniques emerge and tasks lose their problematic character: tasks become routinized, the means to perfect techniques. The theoretical environment takes techniques into account—their functioning and their limits. Then the techniques themselves become routinized to ensure the production of results useful to mathematical activity. Techniques play an essential role in this dynamic. The term "technique" needs to be taken in a wide sense. A technique is a way of solving a task that often presents itself as a complex assemblage of routine parts as well as of conscious reasoning. Thus, technique has a pragmatic role that permits the production of results; but it also plays an *epistemic* [italics added] role (Rabardel & Samurçay, 2001) in that it constitutes understanding of objects and is the source of new questions. (p. 163; trans. Kieran & Guzmán)

In the mid-1990s, in France, when Computer Algebra Systems (CAS) started to make their appearance in high school mathematics classes, researchers (Artigue, Defouad, Duperier, Juge, & Lagrange, 1998) noticed that teachers were emphasizing the conceptual dimensions while neglecting the role of the technical work in algebra learning. However, this emphasis on conceptual work was producing neither a clear lightening of the technical aspects of the work nor a definite enhancement of students' conceptual reflection (Lagrange, 1996). From their observations, the research team came to think of techniques as a link between tasks and conceptual reflection. Lagrange (2003) has argued further that the epistemic role played by technique is particularly important during its period of elaboration. It is exactly this epistemic role played by techniques during their elaboration that serves as a focus for our study of students' developing mathematical theory with respect to factors, multiples, and divisibility.

Our focus on techniques—and their crucial role in the evolution of theory—cannot, however, be separated from the tasks themselves and the function that they play. Lagrange (1999) suggested that task situations ought to be created in such a way as to "bring about a better comprehension of mathematical content" (p. 63) via the progressive acquisition of techniques in the achievement of a solution to the task. Guin and Trouche (1999) added that situations should aim at fostering experimental work (investigation and anticipation). Hoyles (2002) emphasized that a focus on the design of task situations is at the heart of the "transformative potential of [technological] tools in activities" and that, with this focus, "knowledge and epistemology are brought back to center stage" (p. 284).

Mathematics educators have, for decades, tended to view conceptual and procedural knowledge as two quite different entities—the latter being clearly associated with the development of techniques, skills, and mathematical processes; the former, with the understanding of the objects being manipulated (see, e.g., Hiebert, 1986). The notion that a learner could be fostering conceptual learning at the same time as expanding his or her expertise with mathematical procedures has, until now, never really been elaborated.[2] However, the growth of the theoretical perspective that integrates technique, task, and theory, in concert with the use of technology in the learning of mathematics, currently furnishes us with the tools to think about conceptual and technical work in terms of their interactive coemergence rather than as completely separate endeavors.

THE STUDY

The study was carried out in both Mexico and Canada during two consecutive school years. Classes of students from the three grade levels of secondary 1, 2, and 3 participated both years. All classroom sessions were videotaped. During the first phase, that is, during the first year, we observed and analyzed the global tendencies in the evolution of students' techniques. During the second phase of the project the following year, we repeated the study with additional classes of students, but this time, focused on the evolution of techniques within individual students. We modified our methodology so as to obtain the trace of individual student work with the calculator and thereby track the students' techniques as they were evolving. The classroom view screen (the large classroom screen on which was projected the screen of the calculator being used by an individual student) was indispensable for this aspect of the research in that it permitted us to have a video record of all the attempts used by selected students in

[2]Despite theoretical advances made with respect to process–object distinctions (sometimes called operational–structural conceptions), there remain fundamental differences between these distinctions and the procedural–conceptual/technical–theoretical perspectives referred to earlier (see, e.g., Dubinsky, 1991; Dubinsky & Harel, 1992; Noss & Hoyles, 1996; Sfard, 1989, 1991, 1992).

reaching a solution to a given task. In this regard, individual students were regularly invited to come forward to the classroom view screen and to work out a couple of problems on the calculator that was hooked up to it. It is important to note that the students whose work was being recorded on the view screen had not already arrived at a solution beforehand; that is, they were not simply presenting an obtained solution, but were in fact allowing us to witness all the false starts, dead ends, and varied techniques at their disposal in an attempt to find a solution. For the second phase, we also changed calculators (from the TI-83 Plus to the TI-73) so that students would have more information displayed on their calculator screens.[3] Supplementary task questions were included that asked students to describe the techniques or strategies that they were using and were finding effective. We also added to the second phase of the study a series of individual interviews during which we likewise videotaped each student's calculator work as displayed on the view screen.

The aims of the study were to investigate the emergence and development of numerical thinking in students during their first 3 years of secondary school, and the role that calculating technology can play in the evolution of this thinking. Activities were designed that focused on the mathematical content of factors, multiples, and divisibility, involving both prime and composite numbers. One of these activities was a week-long sequence of tasks based on the "Five Steps to Zero" problem (Williams & Stephens, 1992) shown later. Students were initially pretested on their knowledge of the related mathematical material. During the week's activities, there were occasions on which students were questioned about the approaches they were using, and were sometimes asked to describe them to the class, but no classroom teaching as such took place. The teacher may have remarked that a given explanation was not quite precise enough and that the student should rework it so that it might be clearer, but that was the extent of the teachers' feedback during the activities that are described in this chapter. At the conclusion of the week's tasks, 4 students from each of the participating classes were individually interviewed.

The data sources included individual students' worksheets, transcripts of classroom videotapes, transcripts of individual students' calculator attempts using the classroom view screen (we refer to these calculator transcripts as the "numeratim"—a complement to the verbatim of the study), and transcripts of the videotaped interviews that included the numeratim of all calculator work.

---

[3]The TI-83 Plus displays an operation on one line and the result on the next line. In contrast, the TI-73 displays both the operation and its result on the same line (unless, of course, there are too many characters), thus providing the student with a view of twice as many operations and results than are possible with the TI-83 Plus.

TASKS

The Main Problem Situation

The situation that formed the basis for the accompanying tasks was the "Five Steps to Zero" problem (see Fig. 8.1; adapted from a problem by Williams & Stephens, 1992).

Successfully tackling this problem, with the constraint of using only the whole numbers from 1 to 9, involves developing techniques for converting numbers (prime or composite) into other numbers in the same neighborhood (not more than 9 away from the given number), which have divisors that are as large as possible (but not larger than 9), so as to reach zero in five or fewer steps. In the example illustrated in Fig. 8.1, the given number of 151—a prime—needs first to be converted into a suitable composite, with appropriate divisors. Although a variety of composites and their divisors could be used to bring 151 down to zero, students were encouraged to use as few steps as possible. Thus, this problem situation provided fertile ground for learning that, for example, if a number has both a and b as divisors, then it is also divisible by ab.

But awarenesses such as these are not so obvious to students. Grossi (1985, cited in Vergnaud, 1988) has, for example, shown the difficulty that students have in extracting a binary relationship (12 is a multiple of 3) from a ternary relationship ($3 \times 4 = 12$). She observed that, in order "to check whether P is a multiple of M, or whether M is a divisor of P, students divide P by M; but when they find that $M \times N = P$, they are not immediately able to say that P is a multiple of N" (p. 159). In another study of simple number theoretic concepts with preservice elementary school teachers, Zazkis and Campbell (1996) found that, when participants were presented with tasks such as, "In $2 \times 7 \times 13 = 182$, can you say whether 182 is divisible by 7," many could not. Similarly, when asked how the calculator could help them determine the remainder in dividing 10,561 by 24, only 7 of the 21 participants suggested dividing, followed by taking the whole number part of the quotient and multiplying it by 24 to see how many units distant the product was from 10,561 (Campbell, 1996; see also Campbell, 2002).

Take any whole number from 1 to 1,000 and try to get it down to zero in five steps or fewer, using only the whole numbers 1 to 9 and the four basic operations +, -, x, ÷. You may use the same number in your operations more than once, but you must not write more than one operation per line.

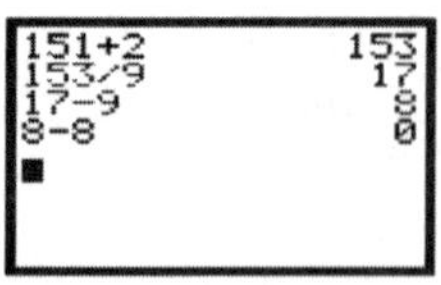

FIG. 8.1. The main situation, "five steps to zero," accompanied by an example (151), displayed on the multiline screen of the calculator.

The findings of Zazkis and Campbell (1996) led them to argue that, "developing a conceptual understanding of divisibility and factorization is essential in the development of conceptual understanding of the multiplicative structure of numbers" (p. 563) and that it should be happening in the early years of secondary school. We were thus interested in seeing how the early secondary students of our study would respond to the situations they would be encountering, and the form that their response would take with respect to the number theoretic elements of factors, multiples, and divisibility.

The Tasks Designed to Accompany the Main Problem Situation

The following set of 10 tasks was developed to accompany the main problem situation.

1. Take the number 144. Write as many ways as you can for bringing 144 to zero, using as few steps as possible.
2. Take the number 151. Write as many ways as you can for bringing 151 to zero, using as few steps as possible.
3. Take the number 732. Write as many ways as you can for bringing 732 to zero, using as few steps as possible.
4. Describe your strategies[4] for minimizing the number of steps.
5. Here is a solution proposed by a pupil for bringing 432 to zero: 432/2 = 216; 216/2 = 108; 108/2 = 54; 54/3 = 18; 18/3 = 6; 6-6 = 0. Show a way of bringing 432 to zero in fewer steps. Explain your strategy. Do you think it will always work? Why?
6. Here is a strategy proposed by a pupil for bringing 731 to zero: 731 + 7 = 738; 738/9 = 82; 82 – 1 = 81; 81/9 = 9; 9 – 9 = 0. Show a way of bringing 731 to zero with fewer steps. Explain your strategy.
7. The number 266 has as its divisors 2, 7, and 19. In other words, 266 = 2 × 7 × 19. What is the best strategy for bringing 266 to zero? Why? Explain why your strategy is best.
8. Here is the strategy proposed by a pupil for bringing 499 to zero: 499 + 1 = 500; 500/5 = 100; 100/5 = 20; 20/5 = 4; 4 – 4 = 0. Show a way of bringing 499 to zero in fewer steps. Explain your strategy.
9. What do you consider to be the best strategies for bringing numbers down to zero?
10. Think of a number less than 1,000 that your classmates would find difficult to bring to zero in five or fewer steps. Write down why you

[4]The term *strategy* was used throughout the set of tasks that were presented to students as it was considered a more general phrasing than "technique" and one that could occasion a broader range of student responses.

think it would be a hard number. Show the solution you found for your "hard number." (This particular task formed the basis for a competition. The class was split into three groups. Each group decided on a number that they considered to be a "hard number" to bring to zero in five steps. Three rounds of competition by two groups at a time trying to solve the hard number of the third group then took place.)

How the Tasks Were Introduced to the Students

The individual classroom mathematics teachers introduced the main problem situation as follows. They began with the example of 360 and illustrated with the view screen that they could get this number down to zero by means of the following operations: 360 / 2 = 180; 180 / 2 = 90; 90 / 3 = 30, 30 / 6 = 5; 5—5 = 0. Without saying any more, the teachers then requested volunteers to come forward to show how they might get to zero in different ways or in fewer than five steps. Teachers were very careful to be neutral with respect to whatever the respondents were proposing, not wanting to foil students' spontaneous techniques with any of the upcoming tasks. After that, students were invited to suggest their own starting numbers, which other students came forward to solve. This was a time dedicated to helping students get a sense of what the basic problem situation was all about. Students then began to work on their task sheets, either individually or in pairs, according to their own preferences, but each student filled in his/her own task sheets.

Types of Calculators Used in This Study

The calculators used in this study were the TI-83 Plus during the first phase and the TI-73 during the second phase. Although these calculators are graphing calculators, their graphing capabilities were not used. They were chosen because of their multiline screens, in contrast with the small one-line screens of simple four-operation calculators. These multiline screens made it possible for students to record and to observe all of the various steps of several calculations at a time. Even though the students had worked with these calculators before, the younger students, in particular, had not had extensive experience with them.

It should also be mentioned that the inherent constraints of these calculators with respect to number theoretic work were capitalized on for our study. They do not provide the complete factorization of a number. Thus, one cannot tell at a glance whether a given whole number is a multiple of some other number, or even whether or not it is prime. As will be seen, these constraints allowed for the emergence of structural ideas about whole numbers, their multiples and divisors, that might not have been possible with more powerful calculators.

A Priori Mathematical Analysis of Techniques Applicable to the Main Problem Situation

The principal question related to the main problem situation concerns the ways in which one might bring a given whole number, less than or equal to 1,000, down to zero, using only the four basic operations, and with as few steps as possible. There are several mathematical techniques that could be employed, for example, prime factorization, but the rule of the situation requiring the use of only the numbers from 2 to 9 within the basic operations made prime factorization an unlikely choice for most cases. By means of a sample task, that of bringing 989 to zero, we examine some of the techniques that could be used, including a variant of prime factorization.

Prime Factorization. According to the *fundamental theorem of arithmetic*, every integer $n \geq 2$ is either a prime or a product of primes, and the product is unique apart from the order in which the factors appear. The prime factors of 989 are 23 and 43; however, neither of these is admissible as a divisor according to the rules of the "five steps to zero" situation. Knowledge of these prime factors could, nevertheless, be used to make estimates of factors that would be permissible—accompanied by a certain amount of trial-and-error adjustment in order to arrive at a number in the vicinity of ± 9 of 989. Versions of this modification of the prime factorization technique are offered later, under the technique called "chaining factors ...."

Criterion for Divisibility by 9. According to the criterion for divisibility by nine, 989 is not itself divisible by 9 because its digits do not sum to a multiple of nine.[5] However, using this criterion will enable us to find two numbers within the ± 9 vicinity of 989 that are divisible by 9. As the digits of 989 sum to 26, adding 1 to or subtracting 8 from 989 will yield the two closest multiples of 9: 990 and 981. Figures 8.2a and 8.2b illustrate bringing 989 to zero by means of the technique involving the criterion for divisibility by 9. As can be seen, however, both versions require six steps to reach zero.

Trial Divisors. Anyone considering the division of the given number 989 by trial divisors would similarly not meet with success because 989 is not divisible by any of the numbers from 2 to 9 without producing a decimal quotient.

[5]The criterion for divisibility by 9 suggests that a number is divisible by 9 if and only if the sum of its digits is divisible by 9. This can be seen by representing a three-digit number N as N = 100a + 10b + c and manipulating this representation to N = (99a + a) + (9b + b) + c = (99a + 9b) + (a + b + c). Because divisibility by 9 of (99a + 9b) is obvious, the divisibility of N by 9 depends on the divisibility of the sum of its digits (a + b + c). This technique is easily extended for any number of digits.

989 +1	990
990 / 9	110
110 – 2	108
108 / 9	12
12 / 6	2
2 – 2	0

(a)

989 – 8	981
981 / 9	109
109 – 1	108
108 / 9	12
12 / 6	2
2 – 2	0

(b)

FIG. 8.2. Illustration of a "solution" requiring six steps to zero.

Chaining Factors to Arrive at a Product Within ± 9 of the Given Number. Although none of the integers from 2 to 9 divide 989, several numbers within ±9 of 989 have such divisors. Knowing that 989 = 23 × 43, when decomposed into prime factors, can be used to make appropriate estimates of composite factors whose product is a number close to 989. Although a first try involving, say, 20 × 45, would be too low, eventually one's trial factors might lead to 20 × 49 (i.e., 5 × 4 × 7 × 7 = 980), which is within 9 of the given number 989.

However, one does not need to know the prime factors of 989 in order to chain factors. A variant of this approach could involve beginning with the given number and carrying out trial division with a chain of composite divisors. For example, one could divide 989 by 81 (i.e., 9 × 9) and notice that the quotient is 12.2. Taking the whole number part of the quotient, 12, and multiplying it by 81, yields a number, 972, that is close to 989, but not close enough. In the same way, one could try 989 / 72 (i.e., 9 × 8) to obtain 13.73, which in turn could suggest 14 × 72 (which, however, equals 1,008). Trying 989 / 49 (i.e., 7 × 7) yields the quotient of 20.18, which when truncated and remultiplied by 49 produces 980. Thus, one obtains a set of factors (7 × 7 × 5 × 4) that results in the product of 980, which is within ±9 of 989, and then suggests a five-step solution by using three of these four factors as divisors (note that the fourth factor is used as the subtrahend in the final step): 989 – 9 = 980, 980 / 7 = 140, 140 / 7 = 20, 20 / 5 = 4, 4 – 4 = 0.

The Division Algorithm. As seen earlier, attempting to bring 989 to zero by means of the technique of dividing by 9 (twice) requires six steps. This can be explained if we look at the decomposition of a whole number as the product of two factors plus a residue. According to the division algorithm, any whole number can be expressed as the product of two whole numbers plus remainder, that is, "For any $b > 0$ and a, there exist unique integers c and d with $0 \leq d < b$ such that $a = b \times c + d$." Thus, the number 989 can be written as $989 = 9 \times 109 + 8$. According to this decomposition, it is clear that when we subtract 8 from 989, we obtain a multiple of 9; but the quotient on dividing 981 by 9 (109) is not a multiple of 9 ($109 = 9 \times 12 + 1$). Thus we need to perform a second adjustment, this time on 109, to obtain another multiple of 9 (109 – 1). These two adjustments and their accompanying divisions by 9 bring us to 12, which then requires another two steps to reach zero (by, for example, successive subtractions). Thus, for the whole numbers from 1 to 1,000, the technique of dividing by 9, the largest of the permissible divisors, is no guarantee of reaching zero in five or fewer steps.

In fact, there are two numbers between 1 and 1,000 for which six is the minimum number of steps in which one can reach zero: 851 and 853. By means of a computer program[6] that generated the minimum number of steps for reaching zero for each of the numbers from 1 to 1,000, it was shown that it takes a minimum of five steps to bring to zero the nine numbers below 851, the nine numbers above 853, as well as 852. The number 851 has as its factors, 23 and 37; and 853 is prime. Bringing either of these to zero would require converting them to one of their nearby neighbors, and thereby necessitate a total of six steps to reach zero.

PHASE 1 RESULTS

Students' Techniques and Their Evolution

The results we present from the first phase of this research study are derived from a comparison of the techniques students used with two of the tasks; one from the beginning of the week's activities and the other from the latter part of the week (the "hard number" task). The choice of these two tasks allows us not only to sharpen the focus of our brief presentation of the Phase 1 results, but also to motivate the second phase of the study. Data are drawn from an analysis of students' task sheet responses, supplemented by the videotapes of the whole class sessions during the week of observations. Five main techniques were found to emerge, some of which were more prominent in the earlier tasks, and others in the later tasks (see Table 8.1). A global portrait of the ways in which students from three classes of one school spontaneously shifted in their use of these techniques from beginning to end of the week's activities is provided in Table 8.2.

That students' techniques evolved without explicit instruction should come as no surprise. Ruthven (1992) has, for example, pointed out that

[6]This computer program was developed by Eric Bainville, using the Maple programming language, at the Leibniz Laboratory of the Institut IMAG, Grenoble, France, in 1999. Unfortunately, it is no longer available.

TABLE 8.1

Techniques That Emerged During Phase 1 of the Research

T1: If the given number is even, divide by an even number such as 2 (or other even numbers, as determined by trial-and-error); if the given number is odd, subtract 9 to make it even and then divide by an even number.

T2: Adjust the given number by addition or subtraction so that it ends in 0 or 5, and then divide by 5.

T3: Adjust the given number by addition or subtraction so as to obtain a number of the form *abc* where *ab* and *c* (or *a* and *bc*) are multiples of some number from 2 to 9; then divide by that number.

T4: Divide by the largest divisor possible from 2 to 9 (i.e., try with 9, then 8, then 7, and so on).

T5: Adjust the given number by addition or subtraction to obtain a multiple of 9 and then divide by 9.

TABLE 8.2

Evolution of Students' Techniques by Grade Level (From Phase 1 of the Study)

Techniques	*Secondary 1 (13 to 14 years)*		*Secondary 2 (14 to 15 years)*		*Secondary 3 (15 to 16 years)*	
	With 151 (n = *21)*	*With the "hard number"* (n = *19)*	*With 151* (n = *25)*	*With the "hard number"* (n = *24)*	*With 151* (n = *15)*	*With the "hard number"* (n = *18)*
T1	3 (14.28%)	1 (5.26%)	0	1 (4.16%)	2 (13.33%)	0
T2	9 (42.86%)	8 (42.10%)	17 (68%)	10 (41.67%)	6 (40%)	5 (27.77%)
T3	2 (9.52%)	6 (31.57%)	5 (20%)	4 (16.67%)	1 (6.66%)	3 (16.66%)
T4	0	0	0	2 (8.33%)	1 (6.66%)	1 (5.55%)
T5	1 (4.76%)	1 (5.26%)	3 (12%)	6 (25%)	0	9 (50%)
Other or none	6 (28.57%)	3 (15.79%)	0	1 (4.16%)	5 (33.33%)	0

when students start to use a calculator, they see it as simply offering a limited range of automatic procedures for familiar processes. But, according to Ruthven, growing confidence encourages novel approaches, most prominently variation on trial-and-improvement. The calculator allows the students to progressively refine conjectures through feedback. He has stated that, "While trial-and-improvement has limitations, it often gives students a means of tackling a problem which would otherwise be intractable; and as the user becomes increasingly familiar with a particular type of mathematical situation, such judgments tend to become more sophisticated" (p. 31). However, as will become clearer on examination of the results presented in Table 8.2, the novel approaches that students developed were not, in the main, "variations on trial-and-improvement," but were based on seeing new structures in their numerical data.

Analysis of Selected Techniques

Technique T2: Adjust the Given Number by Addition or Subtraction So That it Ends in 0 or 5, and Then Divide by 5. The most commonly used approach across all three grade levels at the outset of the study was that of adjusting the given number by addition or subtraction so that it ended in 0 or 5, and then dividing by 5. This is illustrated by the work of Suzanne, from Secondary 1, with the number 151 (see Figs. 8.3a and 8.3b):

As can be seen from Table 8.2, the use of this technique came to be replaced by other more powerful techniques throughout the course of the week. The decrease in its use was most prominent among Secondary 2 students, some of whom shifted toward Technique T4: Divide by the largest divisor possible from 2 to 9, and others to Technique T5: Adjust the given number by addition or subtraction to obtain a multiple of 9 and then divide by 9. The movement to T5 was especially strong among Secondary 3 students. More will be said about this latter technique shortly. At this moment, however, we wish to look more closely at Technique T3, a technique developed, in particular, by Secondary 1 students.

T3: Adjust the Given Number by Addition or Subtraction So as to Obtain a Number of the Form abc[7] Where ab and c (or a and bc) Are Multiples of Some Number from 2 to 9; Then Divide by That Number. This technique merits attention because it may be seen as a transition toward more powerful

151 – 1	=	150
150 / 5	=	30
30 / 5	=	6
6 – 6	=	0

(a)

151 + 4	=	155
155 / 5	=	31
31 – 1	=	30
30 / 5	=	6
6 – 6	=	0

(b)

FIG. 8.3. Work of Suzanne.

[7]We use abc or ab or bc to indicate the digits of a number and to distinguish from *abc*, which, according to algebraic convention, is interpreted as a product.

techniques involving the search for numbers with divisors such as 6, 7, 8, or 9. Its development allowed pupils to work with divisors that were higher than 5, the divisor they had used with Technique T2. Illustrations of the use of the Technique T3 are provided by the attempts of Mariana V and Mariana P (both from Secondary 1) with the numbers given to them in the "hard number" task: 891 and 823 (see Figs. 8.4a, 8.4b).

We note that Mariana V added 9 to the given number and thus obtained 900, clearly a multiple of 9. The theorem-in-action—if we might use Vergnaud's (1988) terminology—that seemed to be at work here can be stated as: "A number that can be expressed as $\underline{a00}$ is a multiple of a." In the case of Mariana P, subtracting 4 from the given number to obtain 819 yielded a number composed of the two blocks 81 and 9—both of which are multiples of 9. The theorem-in-action that Mariana P may have been using is equivalent to: "If the number composed of two digits $\underline{ab}$ is a multiple of c, then the number $\underline{abc}$ composed of three digits is also a multiple of c." We offer an additional example drawn from the work of a Secondary 2 student, Dora, who added 5 to 931 to obtain 936. Her approach suggested the underlying rule: "$\underline{9bc}$ is divisible by 9 if $\underline{bc}$ is divisible by 9." Other forms of the T3 technique that we witnessed included: "If p divides $\underline{ab}$, then p divides $\underline{ab0}$" and "if p divides $\underline{ab}$ and p divides c, then p divides $\underline{abc}$."

Although this technique was obviously limited to those numbers that could be decomposed visually into known multiples, it represented a giant leap forward when compared to their earlier techniques of converting numbers into multiples of 5 (T2) or into numbers that were even so as to divide by 2 or possibly by larger even numbers (T1). It allowed students to begin to consider division by 9—the highest admissible divisor, according to the rules of the situation—as a technique. However, it was at the Secondary 3 level where the greatest number of students evolved toward attempting division by 9 (T5: Adjust the given number by addition or subtraction to obtain a multiple of 9 and then divide by 9) for most of the given numbers. Whereas some of them seemed to be using the criterion for divisibility by 9, it was not so obvious how others were arriving at multiples of 9. We thus decided to revise our observation strategies and data collection methods during the second phase of the study (as described earlier) in order to be in a

891
891 + 9 = 900
900 / 9 = 100
...

(a)

FIG. 8.4a. Work of Mariana V.

823
823 − 4 = 819
819 / 9 = 91
...

(b)

FIG. 8.4b. Work of Mariana P.

position to shed more light on the development of students' techniques for arriving at multiples of 9.

PHASE 2 RESULTS: TECHNIQUES THAT EMERGED

At the outset of Phase 2, the new classes of students began with the same simple techniques as were observed during Phase 1, that is, dividing by 5 if the number ended in 5 or 0, or dividing by 2 if the number was even. However, their techniques evolved, just as was the case with the Phase 1 students. The shifts in methodological approach that we instituted between Phase 1 and Phase 2 did indeed allow us to better capture the details of the techniques toward which pupils evolved, techniques that had remained obscure during the analysis of the first phase. They are presented in the following paragraphs.

Search for a Large Divisor by Trial-and-Error

Students soon realized that division by large numbers could bring them to their goal of zero much more rapidly than dividing successively by 5 or by 2, on condition that these large numbers did indeed divide the given numbers. Thus it was that several students evolved toward the technique of taking the given number and immediately dividing it by 9, then 8, then 7, and so on, with the aim of seeing whether it was divisible by any of these allowable numbers. Figure 8.5 illustrates the work of one student, David, with this technique applied to the given number 546 (note that the last few steps, which do not entail the technique under discussion, have been omitted).

As is suggested by the numeratim shown in Fig. 8.5, David's technique seemed to be functioning as expected (or as he had hoped it might)—L1 to L3 forming one block, and L4 to L6 forming another block. Combining L3 and L6, followed by two subtractions, would clearly produce a four-step solution. It is interesting to note that David skipped over the step of dividing 78 by 7 (L5–L6), likely recognizing that 7 is not a divisor of 78.

The Search for Multiples of 9

The second general technique that we were able to observe in great detail as a result of the second phase of the study concerned the search for multiples of 9. We found several variants of this general technique. It is noted, however, that very few of these variants were used systematically from the beginning of a solution attempt to the end. They were tempered by the need to arrive at zero in five or fewer steps. As dividing by 9 usually involved two steps—one being the adjustment step in order to arrive at a multiple of 9 and the other, dividing that multiple by 9—it was rare that carrying out the technique of dividing by 9 would allow one to arrive at zero in the required

L1: 546 / 9	60.66
L2: 546 / 8	68.25
L3: 546 / 7	78
L4: ans / 9	8.66
L5: 78 / 8	9.75
L6: 78 / 6	13 (1)

FIG. 8.5. Work of David. Note: Legend for calculator-screen transcriptions (i.e., the numeratim): Ln—refers to the current line of the student's work on the calculator display screen. Screen lines that have a bar across them denote lines (or characters) that the student has deleted from the screen. The small number in parentheses indicates the time taken by the student before entering the number or operation that follows—5 blinks of the screen cursor being equal to one unit in parentheses.

number of steps, unless only one adjustment was required—the first one. Thus, the first couple of steps of students' solutions tended to manifest their principal technique, whereas subsequent actions were marked by alternative techniques that could take fewer steps and thus help them reach their goal of zero in five or fewer steps. This integration of two or more techniques is seen in the following extracts, but nowhere more dramatically than in the case of Pablo, which follows. We now present the main variants of the techniques involving the search for multiples of 9.

Division of Given Number by 9, Then Extraction of Whole-Number Part of the Quotient, Followed by Multiplication[8]

(1) Division by 9 Throughout, With or Without Multiplication. Figure 8.6a illustrates the initial work of Pablo with the number 731, followed by his description of the technique used (in Fig. 8.6b).

The technique described by Pablo in Fig. 8.6b lent itself uniquely well to 729 (9 × 9 × 9), the better of the two multiples of 9 close to 731. However, an obstacle was met when he subsequently tackled the number 931 (see Fig. 8.7). But, the obstacle likely had less to do with the fact that the result upon division by 9, that is, 103, was not immediately divisible by 9, than with the fact that reapplication of the technique from its first principles would lead to a solution with more than five steps (note the very long pause indicated at the end of line 4 in Fig. 8.7 after Pablo had arrived at 103).

[8]Although the title of this section captures the principal technique developed by the given student, a second example of his work is included so as to show how students were able to shift from their preferred techniques to meet the demands of specific tasks.

L1: 731 / 9	81.222
L2: 9 x 81	729
L3: 731 – 2	729
L4: 729 / 9	81
L5: 81 / 9	9
L6: 9 – 9	0

(a)

Technique: I divide the given number by 9. With the whole number that I obtain, I do a multiplication by 9. After that, I subtract so as to obtain the multiple of 9 and then I divide by 9 several times.

(b)

FIG. 8.6. Work of Pablo.

L1: 931 / 9	103.44
L2: 9 × 103	927
L3: 931 – 4	927
L4: ans / 9	103 [5]

FIG. 8.7. Pablo's work with 931 begins.

(2) Division by 9, Followed by the Search for Other Large Divisors. Pablo thus shifted in midstream from his technique of looking for multiples of 9 to that of searching for large divisors, by trial-and-error, in order to complete his solution (see his work in Fig. 8.8). However, his attempt to see whether 103 had as divisors 7 or 6 did not meet with success (see L7 and L8). This led him to return to his original technique, involving an adjustment on each application (see L11–L14), with the wild hope of arriving at a miraculous solution within five steps.

L5: 931 – 4	927
L6: ans / 9	103
L7: [3] 103 / 7	14.71
L8: 103 / [2] 6	17.16
L9: 931	
	~~[L5, L6, L7, L8, L9]~~
~~L10: 931 – [4] 3~~	~~928~~
L11: 931 – 4	927
L12: 927 / 9	103
L13: 103 + 5	108
L14: ans / 9	13
	~~[L11, L12, L13, L14]~~ [7]

FIG. 8.8. Pablo's work with 931 continues.

(3) Search for Large Divisors, Right From the Start. At this point, Pablo decided to abandon his technique of searching for multiples of 9 and restrict himself to the less efficient—but perhaps more fruitful for the given number, 931—technique involving the search for large divisors of the given number. As is shown in Fig. 8.9, this trial-and-error technique allowed him to reach zero in five steps (L17–L21).

This example of Pablo's problem solving illustrates how students can work with a variety of techniques in attempting to reach a problem solution, and that the more controlled technique of searching for multiples of 9 can often be integrated with more trial-and-error based techniques, such as that of dividing the given number by various trial divisors.

L15: 931 / (3) 7	133
L16: (1) ans / 9	14.77
L17: (2) 931 / 7	133
L18: (2) ans – 5	128
L19: ans / 8	16
L20: ans – 8	8
L21: ans – 8	0

FIG. 8.9. Pablo's work with 931 ends.

Division of Given Number by 9, Followed by Control of Decimal Residue So as to Yield a New Dividend That Is a Multiple of 9

This technique consists of dividing the given number by 9, and then adjusting the dividend (several times, if necessary) according to the magnitude of the decimal part of the quotient obtained, so as to eventually arrive at a multiple of 9. The following is an example of this technique, drawn from the work of David (see Fig. 8.10), with the given number 323.

The work displayed in Fig. 8.10 (L1–L3) shows that David first divided 323 by 9 and obtained a quotient whose decimal part was .88. This suggested to him that he adjust the given number upward so as to avoid a deci-

L1: 323 / 9		35.88
L2: 326 / 9		36.22
L3: 324 / 9		36 (1)
	~~[L 1, L 2, L 3]~~	
L4: 323 + 1		324
L5: ans / 9		36
L6: (1) ans / 9		4
L7: ans – 4		0

FIG. 8.10. Work of David.

mal residue. But he adjusted upward a little too high and obtained, upon division by 9, the quotient 36.22. His final adjustment downward yielded a dividend that was indeed a multiple of 9, and one that was in the right interval surrounding the given number 323. The subsequent steps that were needed to finish the task required no further adjustments.

Multiplication of 9 by Large Trial Factors (Greater Than 9) So as to Arrive at a Multiple of 9 in the Required Interval

This technique, as illustrated by the work of Marianne, involves starting immediately with a trial multiplication of 9 by another number, with the aim of arriving at a product that was close to the given number (see Figs. 8.11a, 8.11b).

The first four lines of Fig. 8.11a suggest that Marianne was carrying out her trial multiplications without being especially mindful of where she was heading (notice the question she posed to her classmates: "What was the number?" to which someone replied, 971). Her 11th try (L11) involved the number 108, which when combined with the factor of 9, yielded a product within the required interval of the given number 971. Once she had the 972 down to 108, she continued her solution path toward zero, but this time with the technique involving the search for divisors of the now-given number, 108.

The Search for Multiples of 9, 8, 7, and so on, by Controlled Multiplication

This technique, developed by Nicolas, had its roots in his attempts with the third task involving the given number 732. As seen in his fourth try with this number (see Fig. 8.12), Nicolas seemed, all of a sudden, to conceive of multiplying $9 \times 9 \times 9$ in an effort to arrive at a product, which would be in the vicinity of 732. He succeeded indeed. But he did more than just arrive at a

L1: 971 (3)	
L2: 9 × 86	774
L3: (2) 9 × 76	684
L4: 9 × 97	873
[What number was it? Someone: 971]	
L5: 9 × (3) 1	
L6: 9 × 99	891 (2)
~~[L1, L2, ..., L6]~~	

(a)

L7: 9 × 105	945
L8: 9 × 110	990
L9: 9 × 107	963
L10: (2) 9 × 106	954
L11: 9 × 108	972
L12: 971 + 1	972
L13: 972 / 9	108
L14: (3) 108 / 6	18
L15: 18 / 9	2
L16: 2 − 2	0

(b)

FIG. 8.11. Work of Marianne.

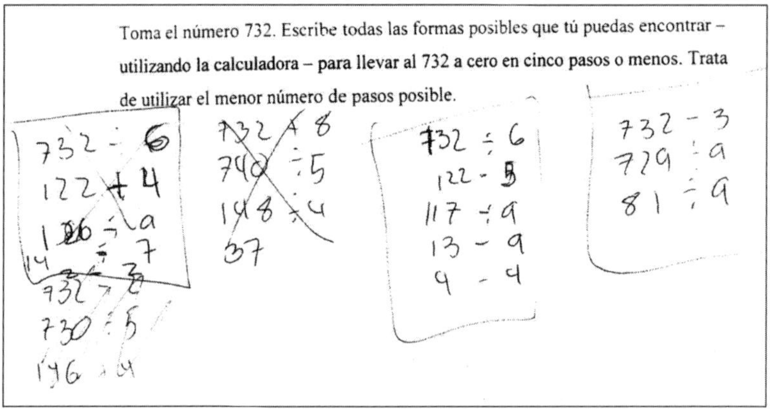

FIG. 8.12. Nicolas's task sheet.

multiple of 9 that was in the vicinity of the given number. By means of these three factors, all of which were admissible in the "five steps to zero" problem situation, he had laid out before him a solution path that would entail using each of these factors as divisors (except of course for the last one, which would be used as a subtrahend).

On the following task sheet, Nicolas described the newly discovered technique as follows: "First, I multiplied 9 × 9 × 9 and I obtained 729; after that I noticed that 732 – 3 was 729 and so I did the same operation, but in reverse."

However, it was only during the individual interview with Nicolas several days later that we were able to observe how he had continued to develop his technique to encompass multiples of numbers other than 9. The following is an extract from that interview (see Fig. 8.13).

This example drawn from the work of Nicolas serves as a contrast to the strictly trial-and-error technique of searching for potential divisors by dividing the given number successively by 9, 8, and so on. Nicolas's technique did not involve testing whether the given number had certain divisors. He started with three trial factors, which he multiplied together with the aim of arriving at a number that was close to the given number. These trial factors were systematically adjusted based on the feedback he received from the various multiplications and their products displayed on the calculator screen. His manipulation of expressions such as 9 × 8 × 6 (see line 38 of Fig. 8.13) suggested an overall awareness that the resulting product was divisible by 9 and by 8 and by 6, and that these factors would yield the set of divisors to be used in bringing the adjusted given number down to zero.

Verbatim	Comments
32. N: Because ... Well, I also have a "technique" that I use. First, I do a multiplication, say, 9x9x3 or something like that to arrive at another number, and I look at that number.	He doesn't really answer the question; rather, he gives another way to attack the problem, using the word "technique."
33. I: Let's see, repeat that for me one more time.	Just to be sure about what he has just said.
34. N: For example, if I have the number 571 and I multiply 9x9, it gives 81	Here he has chosen his own starting number.
35. I: Let us say that I give you the number 431	The interviewer wishes to offer one for which Nicolas may not have a ready-made solution.
36. N: OK, so I go: L1: 9×9 81 L2: ans×3 243 L3: 9×9×4 324 ~~L4: 9~~ L5: 9×9×5 405 So, like that, I arrive more quickly L6: 9×9×5 405	We notice here that Nicolas generates three potential factors, two of them being 9s. Then he systematically adjusts one of them. <u>Legend</u> for calculator screen transcription: see caption in Fig. 8.5.
37. I: But I said 431. With this strategy that you have just described, how do you begin?	An attempt to get him back to the given number 431.
38. N: First, 9x9xsomething, no? Until arriving close to the number. For example, L7: 9×8×6 432	He now controls the last two factors at the same time. The 9 is reduced to 8 while the 5 is increased to 6.
39. I: Yes, I told you 431	Again, to bring the task to its conclusion.
44. N: So, 431 plus 1, divided by 6, divided by 8, and so on.	He states a solution that involves dividing first by the last factor of 9x8x6, and then moving toward the left, inversing each operation, until arriving at 9, which is handled by a final subtraction.
45. I: Let's see.	An invitation to show it on the calculator.
46. N: L8: 431 + 1 432 L9: 432/6 72 L10: ans/8 9 L11: ans − 9 0 And there it is!	He enters his solution into the calculator.

FIG. 8.13. Segment of transcript from the interview with Nicolas.

ANALYSIS OF THE EVOLUTION OF STUDENTS' MATHEMATICAL THEORY

We have, in the two previous sections, seen how students' work with simple approaches of dividing by 2 and by 5 led, first, to the generation of multiples of the form <u>abc</u>, as well as to the strategy of trial division of the given number and, second, to the search for multiples of 9 in the region of the given number. Throughout all of this work, factors and multiples, divisors and divisibility, played key roles. Having presented the main techniques used by students in bringing given whole numbers to zero during the two phases of the study, we now examine these techniques in terms of their number theoretic components.

Students came into this study knowing that even numbers are divisible by 2 and that numbers ending in 0 or 5 are divisible by 5. These bits of theory formed the basis of some of their early techniques. They also knew their multiplication facts, for example, that $4 \times 9 = 36$, and that 36 is thus divisible by 9. Such facts may have formed the basis for the <u>abc</u> technique by which they deduced that numbers such as 936 were divisible by 9. But most three-digit numbers were beyond their range of experience with respect to basic multiplication facts. If they did not want to have to rely completely on trial-and-error to test whether given numbers were divisible by 9, or 8, or 7, and so on, they would have to develop ways of thinking about these larger numbers that would give them some control over the "five steps to zero" tasks that they were engaging in.

The techniques we have described from Phase 2 of the study suggest that the control that students developed was related to the division algorithm. Although they had not been taught this algorithm, students began to think about three-digit numbers in terms of how they might be decomposed into either the product of two factors plus or minus some residual adjustment, or the quotient of dividend and divisor plus or minus a residue. Some students' approaches seemed to be oriented toward multiplication and consideration of factors, whereas for others, the operation of division and consideration of divisors was predominant. Yet, as will be seen, factors and divisors were intimately related in their individual techniques.

Factors Leading to a Multiple of 9

Students, whose work was exemplified by that of Marianne, evolved toward conceptualizing three-digit numbers such as 841 primarily in terms of factors and multiples, plus some residual adjustment, for example, $9 \times 93 + 4$. She would consider 9 as a kind of "fixed factor," and by means of trial multiplication, would search for the "variable factor" such that their multiplication would lead to a good neighboring product, n, that is, one within the required range[9] of the given number, N (as was seen in Fig. 8.11). Then, as soon as she had made the required adjustment to the given number, the fixed factor 9 served as divisor of n. Thus, factors and divisors were closely interrelated in students who, like Marianne, used this technique. Once the two steps of adjustment and division had been carried out, a further analysis was made, this time of the resulting quotient, in order to proceed toward the goal of zero.

Divisors

Students such as Pablo also seemed to think about decomposition in terms of a "fixed factor" and a "variable factor." However, in contrast to Marianne, Pablo did not use repeated trial multiplication (as was seen in Fig. 8.6a) to find the "variable factor." He conceptualized the variable factor as the whole number part of the quotient resulting from the division of the given number

[9]If N were the given whole number, then n is a neighbor of N when $n = N \pm a$, where $a \in \{1,2,3,4,5,6,7,8,9\}$.

N by his fixed factor 9. With the same given number, 841, as was used for the example of Marianne, Pablo would divide 841 by 9 to obtain 93.4 .... The 93 was then multiplied by 9 to yield 837. Thus, Pablo knew at that moment what adjustment had to be made to 841 so as to use 9 as a divisor. Once again, divisors and multiples were closely linked in this "division-first" technique, just as in the "multiplication-first" technique. Note, however, that, whereas Marianne might have had to carry out several trial multiplications before hitting upon an acceptable "variable factor" and "neighboring number n," Pablo was sure to have the right n after just one division and one multiplication.

Multiple Factors Leading to Several Multiples Simultaneously

The technique developed by Nicolas was, in one way, similar to that exemplified by Marianne who thought in terms of 9 as a fixed factor and who, by means of repeated multiplication, sought a single "variable factor" that would yield a product n close to the given number N. However, Nicolas's technique was, in many ways, more powerful. He conceptualized the sought-after "neighboring number n" as the product of several factors, all of which were in the 2 to 9 range. He began with two fixed factors, 9×9, and a trial variable factor, such as 3. Recall his remark during the interview following the week-long class sessions: "First, I do a multiplication, say, $9 \times 9 \times 3$ or something like that to arrive at another number, and I look at that number." Then he would, by means of several trial multiplications involving three factors, manipulate the variable factor—and, if necessary, alter one of his initial fixed factors too—until arriving at a set of three factors whose product was sufficiently close to the given number (as was seen in Fig. 8.13). The huge advantage of his technique was that it gave him an entire solution in one fell swoop. After adding to or subtracting from the given number in order to make a necessary adjustment, Nicolas then used each of his factors (that is, all except one, which was used in the last step subtraction) as successive divisors. This technique also ensured that he had only one adjustment to make during his five-step solution—at the first step, unless, of course, the given number was already a multiple of the three admissible numbers he was using as factors; in this case, there was no adjustment to be made, and the solution thereby involved fewer steps. It is noted, however, that we never witnessed the extension of his technique to four factors for numbers greater than 727. We can only guess that such as extension could easily have become a part of his theoretical thinking, if it were not already so.

Thus, as we have attempted to illustrate—by the work of Marianne, Pablo, and Nicolas described herein—the three concepts of factor, multiple, and divisor constituted a number theoretic ensemble that was conceptualized as a whole. For many of the students of this study, the deployment of one of these concepts implied the use of the other. Lest we create the impression that such theoretical ideas emerged for all the students of our study, let us quickly add that such was not the case.

There were some students who were aware of the criterion for divisibility by 9. These students tended to apply the criterion to the tasks presented to them in a rather mechanical fashion. They seemed not to feel the need to develop any new techniques because they already had one that did the job for them—at least for most of the tasks they were faced with. Their use of the criterion for divisibility by 9 thus deprived them of the rich conceptual learning that occurred for the majority of the other students. Although our study did not focus specifically on those students who knew this criterion, the observations that we were able to make of these students' approaches to their work suggest that mathematics teachers ought to think carefully and cautiously about teaching this criterion to students at these grade levels. Its premature introduction can serve as an obstacle to the further development of their number theoretic thinking.

Other Discoveries Made by Some Students

Students' struggles to find multiples of 9 in the vicinity of the given numbers also disclosed their emergent discoveries of other related ideas, such as, "Within every interval of 9 numbers, there is exactly one number that is a multiple of 9." This theoretical breakthrough for certain students of our study reminded us of the thinking that Zazkis and Campbell (1996) noted among some of the preservice teachers of their research study. Other students in our study whose solution attempts included, "738 / 9 = 82 and 729 / 9 = 81," realized for the first time that, when two adjacent multiples of 9 (e.g., 738 and 729) are divided by 9, the two quotients that are obtained (i.e., 82 and 81) are consecutive. Although these discoveries may seem modest in comparison to some of the techniques described earlier, they should nonetheless be considered significant forward strides, for they prepared the way for further theoretical advances by these students. We close this section by emphasizing that the majority of the students in the study experienced substantial growth in their number theoretic thinking during the week's activities and suggest that tasks similar to those that were employed in this study can clearly be used to good effect in the mathematics programs of students at these grade levels.

THE COEMERGENCE OF TECHNIQUE AND THEORY, AND THE ROLE OF THE CALCULATOR

In concluding this chapter, we return to two issues, the first of which concerns the coemergence of technique and theory. Throughout, we have used the word technique within discussions of the emergence of theoretical knowledge. Why did we find it useful and even necessary to do so? Secondly, what role did the calculator play? How essential was it to the theoretical development that occurred among the students of our study?

Clearly, the students used the "×," "÷," and other operation keys of their calculators in the usual way. But carrying out these operations on the calculator was not the same as carrying them out with paper and pencil. Without having to be preoccupied with the details of the multiplication and division algorithms—details that for years required their attention in paper-and-pencil environments—the students now dealt with tasks where only the product and quotient of these "operations" were displayed. They were thus in an environment permitting the analysis of successive results for potential indications of numerical structure. In fact, faced with all the intermediate results that they obtained on their way to reaching the goal of the task, and that continued to be displayed on the multiline screen of their calculators, students did spontaneously look for relations among the given numbers. Without the distraction of the details of long algorithms, the students developed, with the help of the calculator, techniques that were marked by the search for structural elements in the displayed results—elements of a number theoretic nature.

In this way, a theoretical component was at the core of their techniques—techniques that were rooted in basic mathematical operations but that had been modified by the availability of technology. The fact that the theorizing emerged simultaneously with the evolution of the new techniques suggests that the two dimensions were tightly tied together in this new learning of theirs.

As mentioned earlier, Chevallard (1999) separated technique from the discourse that explains or produces it, but he also admitted that this type of discourse is often integrated into technique—technique that can be characterized in terms of theoretical progress. The techniques developed by Marianne, Nicolas, and others of our study whom we have described, were not just button-pushing techniques. These students were not mindlessly carrying out some routine technique involving a sequence of button presses "to arrive at zero" in a merely pragmatic fashion. No, their techniques played an epistemic role as well, one that, as Lagrange (2002) has remarked, constituted a new understanding of the structure of the numbers they were decomposing. But this epistemic role was achieved within the very action of producing and of interpreting certain phenomena on their calculator screens.

We would be remiss if we did not add that the nature of the tasks was believed to be crucial in this regard. The tasks of this study were designed both to be amenable to a calculator environment and to solicit mathematical theorizing. These tasks would not have been presented in a paper-and-pencil environment where students would have perforce been preoccupied with algorithmic details. Hence our a priori theoretical framework allowed us to both design a study and analyze the data with a view to considering the interweaving of task, technique, and theorizing in a technological environment. Any notion we might have been entertaining with respect to stripping the development of mathematical theory of its robes of task and technique—at least during its emergent period—would have been ill-conceived, at the very least, contrary to the theoretical framework in which this study was conducted. However, the fact that theorizing was constantly intertwined with technique may not be the main message as far as classroom practice is concerned. The point may be rather that theory did emerge with the development of techniques, and that those who limit the use of calculators in their mathematics classes to the simple verification of paper-and-pencil calculations may be missing the even more important contribution of these tools to mathematical learning.

ACKNOWLEDGMENTS

We thank the Social Sciences and Humanities Research Council of Canada (Grant # 410-99-1515) and CONACYT of Mexico (Grant # I32810-S) for their support of the research described in this chapter. We are also extremely grateful to the teachers and students of the two schools that participated in the study.

REFERENCES

Artigue, M. (2002). Learning mathematics in a CAS environment: The genesis of a reflection about instrumentation and the dialectics between technical and conceptual work. *International Journal of Computers for Mathematical Learning, 7,* 245–274.

Artigue, M., Defouad, B., Duperier, M., Juge, G., & Lagrange, J.-B. (1998). *Intégration de calculatrices complexes dans l'enseignement des mathématiques au lycée* [Integration of complex calculators in the teaching of mathematics at the lycée]. Paris: Université Denis Didirot Paris 7, Équipe DIDIREM.

Campbell, S. (1996). On preservice teachers' understandings of division with remainder. In L. Puig & A. Gutiérrez (Eds.), *Proceedings of the 20th Conference of the International Group for the Psychology of Mathematics Education* (Vol. 2, pp. 177–184). Valencia, Spain: PME Program Committee.

Campbell, S. R. (2002). Coming to terms with division: Preservice teachers' understanding. In S. R. Campbell & R. Zazkis (Eds.), *Learning and teaching number theory: Research in cognition and instruction* (pp. 15–40). Westport, CT: Ablex.

Chevallard, Y. (1999). L'analyse des pratiques enseignantes en théorie anthropologique du didactique [The analysis of teaching practices in didactical anthropological theory]. *Recherches en Didactique des Mathématiques, 19,* 221–266.

Dubinsky, E. (1991). Reflective abstraction in advanced mathematical thinking. In D. Tall (Ed.), *Advanced mathematical thinking* (pp. 95–123). Boston: Kluwer.

Dubinsky, E., & Harel, G. (1992). The nature of the process conception of function. In E. Dubinsky & G. Harel (Eds.), *The concept of function: Aspects of epistemology and pedagogy* (MAA Notes, Vol. 25, pp. 85–106). Washington, DC: Mathematical Association of America.

Grossi, E. (1985). *Psychogenèse et apprentissage du concept de multiple* [Psychogenesis and the learning of the concept of multiples] (thesis of the 3rd cycle of French graduate studies). Paris: École des Hautes Études en Sciences Sociales.

Guin, D., & Trouche, L. (1999). The complex process of converting tools into mathematical instruments: The case of calculators. *International Journal of Computers for Mathematical Learning, 3,* 195–227.

Hiebert, J. (Ed.). (1986). *Procedural and conceptual knowledge: The case of mathematics.* Hillsdale, NJ: Lawrence Erlbaum Associates.

Hoyles, C. (2002). From describing to designing mathematical activity. In C. Kieran, E. A. Forman, & A. Sfard (Eds.), *Learning discourse: Discursive approaches to research in mathematics education* (pp. 273–286). Dordrecht, The Netherlands: Kluwer.

Lagrange, J.-B. (1996). Analysing actual use of a computer algebra system in the teaching and learning of mathematics. *International DERIVE Journal, 3,* 91–108.

Lagrange, J.-B. (1999). Complex calculators in the classroom: Theoretical and practical reflections on teaching pre-calculus. *International Journal of Computers for Mathematical Learning, 4,* 51–81.

Lagrange, J.-B. (2000). L'intégration d'instruments informatiques dans l'enseignement: Une approche par les techniques [The integration of instruments of information technology within teaching: A technique-based approach]. *Educational Studies in Mathematics, 43,* 1–30.

Lagrange, J.-B. (2002). Étudier les mathématiques avec les calculatrices symboliques. Quelle place pour les techniques? [Studying mathematics with symbol manipulating calculators. What is the role of techniques?]. In D. Guin & L. Trouche (Eds.), *Calculatrices symboliques. Transformer un outil en un instrument du travail mathématique: Un problème didactique* (pp. 151–185). Grenoble, France: La Pensée Sauvage.

Lagrange, J.-B. (2003). Learning techniques and concepts using CAS: A practical and theoretical reflection. In J. T. Fey (Ed.), *Computer Algebra Systems in secondary school mathematics education* (pp. 269–283). Reston, VA: National Council of Teachers of Mathematics.

Ministrère de l'Éducation du Québec. (1993). *Mathématique 116, Programme d'études.* Québec : Gouvernement du Québec, Ministère de l'Éducation.

Noss, R., & Hoyles, C. (1996). *Windows on mathematical meanings: Learning cultures and computers.* Dordrecht, The Netherlands: Kluwer.

Rabardel, P., & Samurçay, R. (2001). From artifact to instrument mediated learning. *New Challenges to Research on Learning, 22.* Helsinki: University of Helsinki.

Ruthven, K. (1992). Calculator strategies. *Micromath, 8*(1), 31–32.

Secretaría de Educación Pública. (2000). *Programas de estudio, Educación secondaria, Libro para el maestro.* Mexico: Secretaría de Educación Pública.

Sfard, A. (1989). Transition from operational to structural conception: The notion of function revisited. In G. Vergnaud, J. Rogalski, & M. Artigue (Eds.), *Proceedings of the 13th International Conference for the Psychology of Mathematics Education* (Vol. 3, pp. 151–158). Paris: PME Program Committee.

Sfard, A. (1991). On the dual nature of mathematical conceptions: Reflections on processes and objects as different sides of the same coin. *Educational Studies in Mathematics, 22*, 1–36.

Sfard, A. (1992). Operational origins of mathematical objects and the quandary of reification—the case of function. In E. Dubinsky & G. Harel (Eds.), *The concept of function: Aspects of epistemology and pedagogy* (MAA Notes, Vol. 25, pp. 59–84). Washington, DC: Mathematical Association of America.

Vergnaud, G. (1988). Multiplicative structures. In J. Hiebert & M. Behr (Eds.), *Research agenda for mathematics education: Vol. 2. Number concepts and operations in the middle grades* (pp. 141–161). Reston, VA: National Council of Teachers of Mathematics.

Vérillon, P., & Rabardel, P. (1995). Cognition and artifacts: A contribution to the study of thought in relation to instrumented activity. *European Journal of Psychology of Education, 10*, 77–101.

Williams, D., & Stephens, M. (1992). Activity 1: Five steps to zero. In J. T. Fey (Ed.), *Calculators in mathematics education* (Yearbook of the National Council of Teachers of Mathematics, pp. 233–234). Reston, VA: NCTM.

Zazkis, R., & Campbell, S. (1996). Divisibility and multiplicative structure of natural numbers: Preservice teachers' understanding. *Journal for Research in Mathematics Education, 27*, 540–563.

Chapter **9**

Learning Number Theory Concepts Via Geometrical Interactive Computerized Setting

Ilana Lavy
Emek Yezreel Academic College

In my experience, junior high school students often resist attempts to teach them elementary number theory concepts, because they feel they have already learned them in elementary school. Therefore, I tried to teach these concepts through activities in a geometrical setting—the software environment MicroWorlds Project Builder.[1] In this chapter, I describe an indirect engagement of 7th-grade students in elementary number theory concepts. In fact, the students were engaged in geometrical activities in a computerized setting. These activities led to the use of basic number theory concepts such as prime, divisor, and greatest common divisor (GCD).

The description of the activities the students (Noam and Jacob) were engaged with highlights the influence of computer environment on both individual and collaborative learning. These two students involved in this study are a special case of two students that are having significant problems communicating mathematically, and the computer plays a mediating role in helping them overcome these difficulties. By the mediation of the computerized setting, they use screen images—each in a different way—to help them take a meaningful part in the mathematical discourse running be-

[1]MicroWorlds Project Builder is a product of Logo Computer Systems, Inc. (LCSI). For more details see the company's web site: http://www.microworlds.com

tween them that eventually led to the emergence of concepts that are closely related to basic number theory concepts.

The characterization of ways Noam and Jacob look for mathematical regularities, and how they communicate with each other, was made possible due to the learning profiles of the two students as were revealed in a previous study (Lavy & Leron, 2004). In what follows, I show how characteristics of each student contributed to the shared process of the concepts' construction, and how the students used the computerized setting to enable this process. I consider two activities in which Noam and Jacob were engaged. I focus on patterns these students attended to during the course of their activities that gave rise to construction to concepts related to number theory.

THE LEARNING ENVIRONMENT

The computer environment is based on *MicroWorlds Project Builder* (MWPB), which is a logo-based construction environment.

The *Geoboard* is a mathematical package implemented on top of the basic MWPB environment. It consists of a simulated round board with pegs at constant intervals on its perimeter. The students can modify the number of pegs by using the Logo instruction *newboard <input>*. For example, *newboard 12* will create a 12-peg geoboard.

A class of ten 7th-grade students participated in this research. The participants met several times after school hours in the computer laboratory and explored the effects of the instruction *repeat n [jump k]* on geoboards of a varying number of pegs—the "threaded pins" problem (Mason, Burton, & Stacey, 1982, pp. 52–56).

The instruction *jump k* results in a line segment being drawn from the "current" peg to the k-th peg counting clockwise, which now becomes the new current peg. Thus the instruction *repeat n [jump k]* results in the command *jump k* being executed *n* times in succession. Each choice of specific values for *n* and *k* results in one of two distinct patterns; (1) a screen image of a regular polygon or (2) a star with varying number of vertices (Fig. 9.1). Students were encouraged to look for finer grained mathematical patterns and conceptual relations among the input numbers, the shapes, and the number of vertices of the resulting polygons or stars. These investigations eventually led to the emergence in the students' discourse of concepts such as prime number, divisor, and greatest common divisor (GCD).

Although the three numbers appearing in the investigation (the number of pegs, the repeat, and the jump) can be chosen separately, in practice, the students always chose the number of repetitions to be the same as the number of pegs on the board. The result—that the figure was sometimes drawn several times over (for example, twice in Fig. 9.1b)—was ignored by the students most of the time.

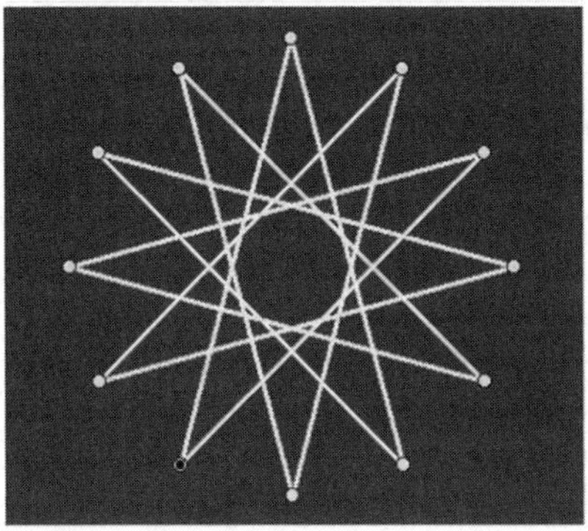

FIG. 9.1a. Twelve-star resulting from the command *repeat 12 [jump 5]*.

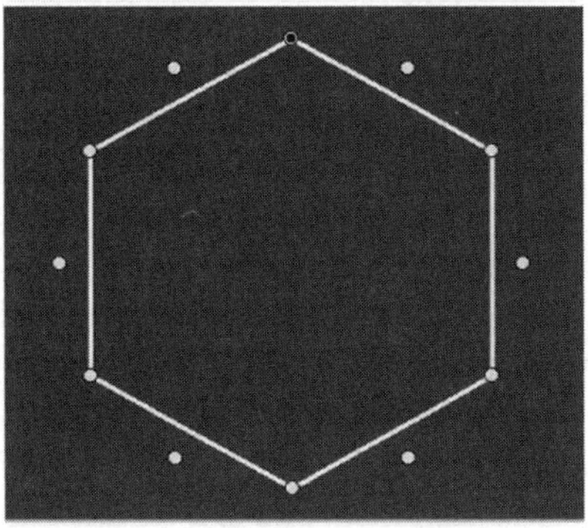

FIG. 9.1b. Six-sided regular polygon resulting from the command *repeat 6 [jump 2]*.

METHODOLOGY

A pilot study preceded the present study, in which 18 seventh-grade students freely explored the geoboard environment. The students represented all talent levels and came from an ordinary school. Because the students were fascinated by the graphical effects of huge inputs, it eventually became a serious distraction and did not facilitate finding interesting mathematical relationships. As a result, I decided to ask the students to use small input numbers and in general to attempt a more systematic investigation. In the two stages of the study, I served as both the teacher and the researcher. However, I kept my intervention to a minimum, limiting it to encouraging students to maintain a focus on the task of investigating the shapes produced by the command *repeat n [jump k]*.

The methodological framework used in this study made use of qualitative research methods of inductive analysis (Goetz & Lecompte, 1984) and discourse analysis (Brown & Yule, 1983; Harre & Gillet, 1994; Sfard & Kieran, 2001). For a more general investigation with this population of students including a detailed description of the employed methodological framework used in this study, see Lavy and Leron (2004).

Initial observation of the verbalized discourse of the students during the various sessions of the exploration process gave the impression that there was no effective collaboration between Noam and Jacob and that each one of them was focused on themselves. However, further in-depth analysis revealed a totally different picture.

The analysis of the verbalized discourse during the investigation sessions was done in two main phases; inductive analysis and interactional analysis.

Inductive analysis was applied in the first phase to look for phenomenological categories. Inductive analysis (Goetz & Lecompte, 1984) is a method that integrates scanning the data while looking for phenomenological categories, with successive refinement of those categories when confronted with new events and interpretations. In keeping with this approach, no predetermined criteria or categories were imposed. This kind of analysis helped in identifying and characterizing different modes of working such as modalities of environment-dependence and a gradual movement toward focusing mainly on the number theory concepts and their attributes.

In the second phase, the discursive tool of interactional analysis (Sfard & Kieran, 2001) was applied in order to characterize learning profiles for the students. Interactonal analysis enabled me to answer questions such as: Who is leading the inquiry process in each stage of the investigation process, and how? Is each one of the students interacting with his mate or speaking mainly to himself? What is the contribution of each one of them in the construction of the concepts? How were they using number theory concepts while trying to explain the connections between the Logo instruction *repeat n [jump k]* and the resulting stars and polygons that appeared on the computer screen, and so forth. Answers to these questions helped me identify and characterize their individual contribution to the construction of concepts that were tightly related to number theory concepts. When encountering utterances that were difficult to comprehend, I applied focal analysis (Sfard, 2000) in an attempt to figure out what the student was seeing or thinking (physically or in his mind) while asserting a particular utterance.

In this study the focus is on one student pair, Noam and Jacob, whose actions, discourse, and screen productions were captured by a video camera. These students were chosen because, more than the other students in class, they tended to "think aloud" during their work. The camera was positioned behind the students, so that it captured the computer screen and the keyboard and often the students' faces, as they turned to talk with one another. The major part of the research data is the discourse of these two students during the activities plus screen pictures at every stage of the activities.

RELATED STUDIES

In this section, I provide snapshots from related studies on classroom discourse, interactive computer environments, and research on understanding elementary number theory. Classroom discourse was discussed broadly by Cazden (1988), which emphasized its special attributes and importance. She emphasized the importance of peer interactions: "The only contexts in which children can reverse interactional roles with the same intellectual contents, giving directions as well as following them, and asking questions as well as answering them, is with their peers" (p 134). She has also related to the role of the computer in peer interactions pointing that the placing of computers in classrooms increases collaboration among peers.

Much has been written on the role of interactive computer environments in facilitating mathematical learning (e.g., diSessa, 2000; Noss & Hoyles, 1996; Papert, 1980). In the present study, the role of the computer environment in facilitating the social interaction between students is emphasized. Particularly, the environment provides students with a language for talking about their mathematical explorations, and it facilitates their involvement in the processes of generalization, abstraction, problem solving, and successive refinement of their mathematical intuitions.

Understanding elementary number theory concepts is fundamental for many branches of mathematics, yet it has received relatively scant attention in the mathematics education research literature especially regarding students of elementary and middle school. Part of the existing studies discusses elementary number theory concepts while investigating other mathematical issues (Leron, 1985; Lester & Mau, 1993; Martin & Harel, 1989; Movshovitz-Hadar & Hadass, 1991). Part of the studies explored the multiplicative structure of numbers through problem solving (Ball, 1990; Graeber, Tirosh, & Glover, 1989; Greer, 1992); and part of the studies explored the understanding of elementary number theory concepts by preservice elementary teachers (Campbell & Zazkis, 2002; Zazkis, 1999; Zazkis & Campbell, 1996a, 1996b). An activity in which the concept of greatest common divisor (GCD) was examined from a geometric standpoint was suggested by Olson (1991), but there was no research found on its impact on the understanding of the GCD concept.

Sinclair, Zazkis, and Liljedahl (2003) explored the impact of the MicroWorld *Number Worlds* on the understanding of preservice elementary teachers of concepts and relationships underlying elementary number theory. They found that interaction with *Number Worlds* had a positive effect in helping participants construct those concepts.

The present study focuses on the learning of basic number theory concepts of seventh-grade students in an indirect manner. Namely, the learning of these concepts is via engagement with geometrical explorations in the geoboard setting.

COOPERATIVE CONSTRUCTION OF SPONTANEOUS MATHEMATICAL CONCEPTS

This section includes a description of a process of how Noam and Jacob helped each other—with the mediation of the interactive geoboard environment—in the construction of spontaneous mathematical concepts. First the students' learning profiles as were revealed in a previous study (Lavy &

Leron, 2004) are introduced. In addition, data is presented to show how the students constructed two working concepts. For reasons that are discussed later, the first of these I refer to herein as *n-star* and the second, the students incorrectly referred to at first as *common denominators*.

The first case focuses on the construction of the n-star concept. I demonstrate how the two students successively refined this concept from a rough first beginning to a relatively elaborate final generalization, and how they gradually connected the n-star concept to elementary number theory concepts (namely, divisor, odd number, and prime number). The second case focuses on the construction of the so-called common denominators concept. I demonstrate a refinement of the investigation process itself, from unplanned trial and error to goal-directed hypothesizing and redressing that term's relation to elementary number theory concepts (namely, common divisors and GCD concepts).

The Students' Learning Profiles as a Contextual Framework for Describing Their Construction of Mathematical Concepts

The learning profiles of Noam and Jacob that were drawn by Lavy and Leron (2004) contributed to the understanding of the construction of the n-star and the common denominators concepts. Namely, these profiles consisted of the characterization of each student regarding both the learning skills he demonstrated during the investigation process and his contribution to the cooperative learning they shared.

Noam. He was the "engine" of the investigation, initiating the next experiment on the computer, and trying to interpret the result, which in turn leads to the next stage, and so forth. Most of the communication he had with Jacob was carried out through the computer images he generated. These screen images were an integral part of his verbal utterances, and the latter cannot be fully understood without the former. Noam is constantly in control of the keyboard and in general of the investigation itself. He is bright and curious, likes to try things out on the computer, but does not stay long with each investigation and tends to jump from one thing to another. His utterances are mostly on the level of concrete examples, even when he is thinking of a general statement. He is self-centered and has great verbal difficulties and his formulations are vague and incomplete.

Jacob. Mostly reactive speech, meaning that he is mainly trying to penetrate Noam's private channel. Because Noam's utterances are incomprehensible in themselves, Jacob relies on the screen images for making sense of them and to play his part in the investigation process. This way he does not drop out and in fact he is usually the first to make a general statement, which is subsequently refined by Noam. Jacob is in many respects the counterpart of Noam. He has a strong need for clarity, systematicity, and structure in trying to understand the situation under study. He is slow in thinking but (or perhaps, hence) has a strong tendency to think systematically and formulate generalizations, often prematurely after too few examples and with faulty conclusions. He is very eager to create the impression that he is actively involved in the investigation and that he understands the situation. It should be mentioned in this connection that each time the teacher approached the pair, Jacob raised his voice in an obvious attempt to show the teacher that he was as actively involved in the investigation as his partner.

Construction of the "n-Star" Concept and Its Relation to Basic Number Theory Concepts

During the first three sessions of the activity, the students became acquainted with the geoboard environment. They investigated the relationship between the jump (k), the number of pegs on the geoboard (n), and the number of vertices in the shape drawn on the screen by the instruction *repeat n [jump k]*. In these sessions they worked mainly with a 24-peg board.

The construction of the n-star concept took place in the fourth session of the activity. The n-star concept emerged from a synthesis of the geoboard learning environment (stars and polygons, the number of pegs n and the jump k) and the related elementary number theory concepts (primes and divisors). The name "n-star" refers to a star of n vertices, where n is the number of pegs on the geoboard. It can be shown (Lavy, 1999) that the command *repeat n [jump k]* will produce an n-star if and only if n and k are relatively prime. The construction of the n-star concept was the result of a process in which the students learned how to generalize from examples, create and refine hypotheses, and find mathematical patterns behind the screen images.

In fact, during their discourse, the students had been using the term *24-star* for what I decided to term as n-star. This decision is based on three "proofs" (informal arguments) that 24-star for them stood for something more general than the particular star on a 24-peg board. First, students used the term *star of all* while working with a 24-peg board (cf. lines[2] 4.179, 4.182, 4.183 below). Second, they used of the term *24-star* while working in boards with *different* numbers of pegs:

> [extract 4.272, while working in a 20-peg board]: "now you are getting a star of this 24"

[2]4.117—the number 4 stands for the session's number and 117 stands for the sequential number of the utterance in this session.

[4.372, while getting a 5-star in a 15-peg board]: "which is not a 24-star, it is a star of ..."

Third, Noam made an explicit general statement while working with a 24-peg board [4.298]: "... it will come out a star with number of vertices like the number of pegs ...."

To demonstrate the process of concept construction by successive refinement, I have chosen a scene from the fourth session, in which the stages are clearly expressed. The following excerpts are translated from the students' speaking language (Hebrew). Clarifications of vague utterances are added in brackets. First, here is the raw data:

[4.177] Noam:	Here I think ... Look I think that ... I want to check something, OK? Let me see, let me see, I want newboard 24, jump ... [moves back to a 24-peg board, types repeat 24 [jump 7] and a 24-star appears on the screen.]
[4.179] Noam:	I think I arrived to some regularity in stars of 24, of the ... all the vertices. Here jump 6 ... here I got it [goes on typing]
[4.180] Jacob:	Look, if you do a prime number, you get a star.
[4.181] Noam:	No, not just a star, a star of 24.
[4.240] Jacob:	(to the teacher): and you put in the jump an odd number, then I get a star. Why? Because it is not divisible.
[4.277] Noam:	... For me a star is only if it is not divisible.
[4.289] Noam:	But look, in 24, but in a prime which is not divisible, it always comes out a star of 24. [pointing at the screen] look. jump 7 is a star of 24.

As was previously mentioned, the development of the concept n-star reveals a process of *concept refinement*. First I explain the previous excerpt and demonstrate the successive refinement of the concept development, and then I discuss its relation to basic number theory concepts.

Line [4.177] indicates that Noam is about to discover the reason for what makes an n-star. The rest of the excerpt documents his struggle (working together with Jacob) to formulate this discovery.

This is the first time in the discourse that Noam and Jacob used terms of elementary number theory concepts like odds and primes while trying to explain the resulting geometrical shapes on the computer screen. At first, the students noticed that "a jump of prime number makes a star" [4.180]; later, "a jump of prime number creates a 24-star" [4.181]; then they are sidetracked for a moment, apparently making the well-known confusion (Zazkis, 1995) between odd and prime numbers, "a jump of odd [prime?] number creates a 24-star because the jump does not divide 24" [4.240], which is eventually confirmed by Noam [4.277]. Finally, they uttered, "a jump of a prime number which is not a divisor of 24, makes an 24-star" [4.289].

In fact, the students did not make one more step to the complete generalization, namely, if *k* and *n* are relatively prime, we get an n-star. One possible reason for this is that they did not encounter any example of a nonprime jump that makes an n-star because they investigated geoboards with small numbers of pegs (for example, there is no number that is smaller than 24, and relatively prime to 24, that is not prime). Figure 9.2 presents the relations, as perceived by students, between elementary number theory concepts and the geometrical shape appearing on the computer screen. More specifically, relations between shapes resulting from the inputs (*n* and *k*) in the instruction *repeat n [jump k]* and concepts, such as odds, divisors and primes.

Constructing the n-star concept was the result of a joint effort by Noam and Jacob. The broken lines refer to Jacob's part and the solid lines refer to Noam's. In the rectangles on the left appears the kind of jump and on the right the resulting shape—a star (that is, a star with the number of vertices less than *n*), or an n-star.

The first rectangle appearing in the left side in Fig. 9.2 refers to Jacob's utterance (4.180) that in an *even geoboard* a jump by a prime number will result in a star. A few minutes later, Jacob formulates utterance (4.240), that it takes an odd number of jumps to get a star. The second rectangle refers to Noam's utterance (4.181) in which he characterizes the star as a special case of a 24-star. This is Noam's discovery and he is the one who first used the term 24-star. This utterance is a refinement of Jacob's previous utterance (4.180), in which he was not yet aware that this was not just any star, but specifically a 24-star. The fourth rectangle on the left side of Fig. 9.2 refers to

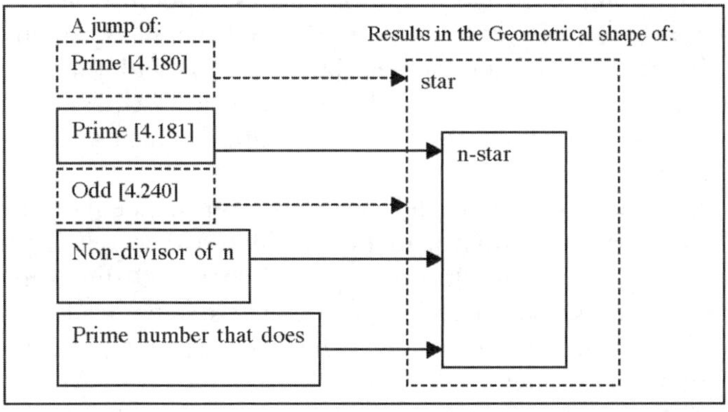

FIG. 9.2. Making connections between number theory concepts and geometrical shapes.

Noam's utterance (4.277) in which he emphasizes the fact that the size of jump is a number that does not divide n. Although Noam used the word *star* in this utterance, looking at the preceding and following parts of the transcript, it is clear that he meant a 24-star. The fifth rectangle refers to Noam's utterance (4.289) in which he connects (4.277) with (4.180) and arrives at the conclusion that in an even geoboard, a jump by a prime number that does not divide 24 creates a 24-star. By the end of this process, the concept 24-star (which, please recall, I am referring to here as n-star) has become a familiar term for these students, used freely in the subsequent stages as if it were part of their natural language.

During the first three sessions, the students were just observing various polygons and stars resulting from the various inputs of n and k in the instruction *repeat n [jump k]*. In the fourth session, they started to see through the geometrical shapes appearing on the computer screen. The interactive nature of the process encouraged the students to try more inputs, which eventually led them to find mathematical connections between certain inputs of n and k and the resulting geometrical shape. Thus, the computer environment played an important role in the process of these students constructing the n-star concept, and their successive refinement thereof. By the immediate response (polygon or star) to each input of n and k, the computer environment enabled and empowered these students to make hypotheses and to verify them.

Observing the discourse of Noam and Jacob, however, also reveals communication difficulties between the two. These difficulties apparently resulted from Noam's inability to express his ideas in a simple and clear manner. In fact, he used incomplete sentences, which required that Jacob try to follow him through the screen images Noam produced. This was possible due to the computerized geoboard being functional in two ways. First, Noam himself completed many of his sentences by pointing at a geometrical shape resulting from certain inputs of n and k that he typed on the keyboard. The resulting shape helped Jacob understand the incomplete sentence uttered by Noam. Second, those geometrical shapes enhanced the students' ability to look for and verify mathematical patterns in the parameters generating them.

These students' work had initially been based on unsystematically changing the values of k and n, which actually contributed to uncovering new regularities. When Noam and Jacob suspected they had discovered some regularity, they would start checking cases more systematically in order to verify it. This exhaustive process would hardly have been possible without the computer environment.

As mentioned earlier, because the intervention of the teacher in the students' learning process was minimal, it might be said that the students, through the mediation of the geoboard environment, had succeeded in developing the n-star concept. Although they did not bring detailed explanations regarding the connection between the attributes of numbers that are relatively prime and the resulting n-star, they were eventually able to predict the inputs of n and k that result in n-stars.

In addition, as can be seen from the discourse during the first three sessions, most of the words used by the students belong to the geoboard environment. From the fourth session on, basic number theory words, such as primes, divisors and so forth, were often used. This implies a shift in their thinking toward seeing the mathematical patterns *behind* the geometrical figures on the computer screen. They successfully developed the ability to find connections between basic number theory concepts and geometrical shapes appearing on the computer screen in the geoboard setting.

Although the students uncovered the mathematical regularity related to the connection between the inputs (n and k) in the command *repeat n [jump k]* and the resulting geometrical shape (star or polygon), in the case of n-star, they did not give clear and detailed justification for it. In the case of an n-peg geoboard, the jth peg represents all the numbers m that satisfy $m \pmod{n} = j$ (or alternatively: $m = j + t \times n$). For example, in the case of a 12-peg geoboard, the first peg represents the numbers 0, 12, 24, 36 …; the second peg (clockwise) represents the numbers 1, 13, 25, 37 …; the third peg represents the numbers 2, 14, 26, 38, 50 … and so forth (Fig. 9.3).

Working with the geoboard helped Noam and Jacob visualize the previous connections, but because they did not see the numbers each peg represents (as presented in Fig. 9.3), they did not explain *why*, in the case that the jump size and the number of pegs on the geoboard are relatively prime, one has to pass through *all* the pegs on of the geoboard. In this case, because n and k are relatively prime, each jump ends in a different peg (representing *one* of the numbers) and $n \times k$ steps are needed to return to the starting peg. Thus the star is completed.

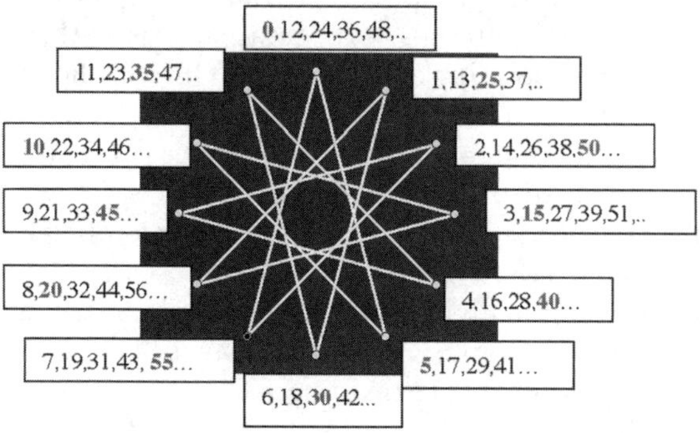

FIG. 9.3. The group of numbers each peg represents in 12-peg geoboard—the case of *repeat 12 [jump 5]*.

Construction of the "Common Denominator" Concept and Its Relation to Common Divisors and the Greatest Common Divisor (GCD)

The second concept developed by the students was the concept of common denominators. This name was used by Noam while he and Jacob were investigating the case in which, for certain inputs of n and k, the resulting star from the instruction *repeat n [jump k]* has less than n vertices. Analysis of their discourse reveals that Noam had a hypothesis in mind regarding the relation between n and k in case of getting a star with less than n vertices and by giving more and more examples (various inputs of n and k), he tried to convince Jacob and himself that his hypothesis was valid.

In fact, the regularity Noam and Jacob were trying to uncover is that $v = n/g$, where v is the number of vertices, and g is the greatest common divisor of n and k. Initially, Noam had only a very vague notion of the underlying mathematical regularity. In practice, he did the division by looking for the multiplicative complement x (which is the desired v) such that $g \times x = n$. It seems that since this "complementing" process reminded Noam of the addition of fractions, he used the term "common denominators" to refer to it (Lavy & Leron, 2004).

The following excerpt is taken from the end of the fourth session, in which Noam and Jacob started to explore stars with less than n vertices. The first geoboard used by Noam is of 15-peg geoboard and the jump size is 10:

[4.416] Noam: I am checking its [the resulting star] relation to the common denominators. Here look for example if I make repeat 15 [jump 10], yeah, 10 is a common denominator of 5, right? Common divisor.

[4.417] Jacob: Of 15.

[4.418] Noam: No, of 5. Here, you see what this is? [types the instruction repeat 15 [jump 10]. A triangle appears on the screen. Noam continues with great surprise. Ah, here, you see why? [pointing at the screen.] This will always come out 10, the common divisor which belongs to it, [with emphasis] their common divisor times what will give it the number of the board. You understand this?

[4.419] Jacob: In short, ... if you do 15 divided by 5 you get 3—so you get a triangle.

[4.420] Noam: [Emotionally] But you understand why it came out a triangle, don't you? 'cause it is 5 times 3!

[4.421] Jacob: Because you are dividing it.

[4.422] Noam: No ...

[4.423] Jacob: So you have 5 between each one [pointing at the triangle on the screen], so you have a triangle.

[4.424] Noam: [executing again repeat 15 (jump 10), and again a triangle appears on the screen]
[4.425] Jacob: 10, it does [referring to the size of the jump he has seen on the screen]
[4.426] Noam: You see, but do you know why this happens?
[4.427] Jacob: No.
[4.428] Noam: Because 3 is what gives to 15 to 5, and really 15 is common denominators. And if you do repeat 15 [jump 5] you get a triangle, you got it?
[4.429] Jacob: [giving up] I got it.
[4.430] Noam: The problem afterwards is to formulate it. But let's check it in another place, for example 20, no 30 [selects a 30-peg geoboard].

In the previous excerpt, Noam is trying to explain to Jacob (and to himself) a new phenomenon that he observed accidentally. When they were exploring the different shapes of a 15-peg geoboard, Noam had expected that a jump of 6 in a 15-peg geoboard to result in a star, but what he got was a 5-star, not the expected 15-star. Because it did not meet his expectations he reacted with: "Oh, I think I start to understand ... look I think I start also to understand something" [4.362].

Jacob, who insisted on being an integral part of the exploration process, but had difficulties understanding Noam's incomplete ideas, followed the shapes on the computer screen. Nevertheless, he did not understand what Noam was doing and asked him for explanations. Noam, on the other hand, concentrated on himself, busy checking jumps of 7 or 8 in a 15-peg geoboard. At this stage of the investigation, he could not yet clearly formulate his ideas, but he was able to produce and check up more examples that result in stars with less than n vertices.

In the previous excerpt, Noam started to examine his conjecture regarding stars with less than n vertices. The first example he chose turned out to be not suitable because instead of getting a star with less than n vertices, as he expected, he got a triangle. Utterance [4.416] is clarifying Noam's state of vagueness in this stage of the inquiry. This utterance starts with Noam's declaration of his intention to find the relation between the resulting star to the common denominators, which is followed by *10 is a common denominator of 5*, and ends with his correction, *common divisor*.

Realizing that Jacob does not understand him, Noam rephrased the utterance [4.418], which is more elaborated. Although they were looking at the same screen image, each one of them interpreted it differently. Jacob gave very clear interpretation: The division result of 15 by 5 is 3, hence the resulting shape is a triangle. Noam, on the other hand, who initiated this example to demonstrate a case in which a star of less than n vertices could be reached, tried to understand why they got a triangle instead of star. Although he saw the same screen image as Jacob, he referred to it as a star with less than n vertices because the jump size (10) has a nontrivial common divisor with 15. Realizing that he could not verbalize his incomplete understanding, he turned to the computer [4.424] and repeated the demonstration of *repeat 15 [jump 10]*, from which Jacob indeed got the meaning of the number 10 here [4.425]. After a few more attempts to explain his partially formed ideas to Jacob, Noam finally said in frustration, "the problem is to formulate it" [4.430].

By the end of the fourth session, Noam's expressions relating to the investigation were so vague that it was almost impossible to figure out what his state of understanding was. At this stage of the inquiry, the following diagram was used (Fig. 9.4) to help illustrate the difference between Noam's and Jacob's respective understandings while looking at the same screen image.

Figure 9.4 includes three columns: The left column refers to the students' utterances, the middle column refers to what the students actually see when they phrase their utterances, and the right column presents my interpretation of the students' utterances. Following the students' utterances, it could be said that although the students saw the same screen image (triangle in a 15-peg geoboard), they actually saw different shapes and attributed to them different properties. Jacob, who followed the investigation process

utterance	**focus**	**interpretation**
[416] **Noam**: I am checking its [the resulting star] relation to the common denominators. Here look for example if I make repeat 15 [jump 10], yeah, 10 is a common denominator of 5, right? Common divisor. [417] **Jacob**: Of 15. [418] **Noam**: No, of 5. Here, you see what this is? [types the instruction repeat 15 [jump 10]. A triangle appears on the screen; Noam continues with great surprise] Ah, here, you see why? [pointing at the screen] This will always come out 10, the common divisor which belongs to it, [with emphasis] their common divisor times what will give it the number of the board. You understand this? [419] **Jacob**: In short, [...] if you do 15 divided by 5 you get 3 – so you get a triangle.	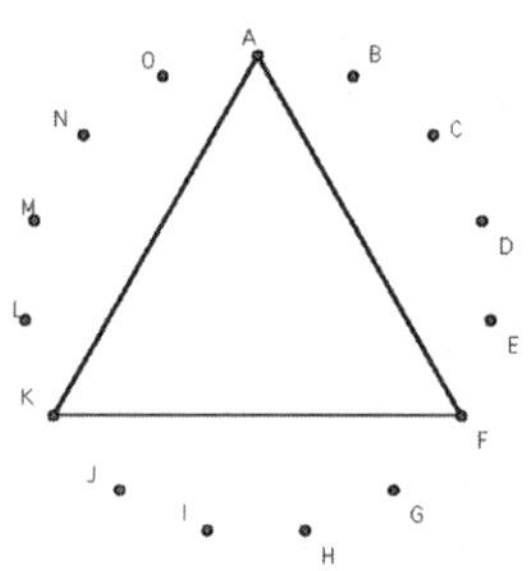	**Noam**: In a 15-peg geoboard, jumps of 5 pegs and jumps of 10 pegs yield in the same shape (triangle) since they have the same common denominator with 15. **Jacob**: In 15-peg geoboard a jump of 5 pegs yields a triangle since 15/5=3

FIG. 9.4. Analysis of 4.416–4.419.

mainly from observing the screen images on the computer screen, saw a simple triangle whereas for Noam, who meant by his example to demonstrate the case in which a star with less than *n* vertices can be reached, the triangle represented a star with less than *n* vertices. From utterance [4.19], it is clear that Jacob divided the 15 pegs (A to O) of the geoboard into groups of five pegs each ((A, B, C, D, E), (F, G, H, I, J), (K, L, M, N, O)) and he got three groups that cover all the geoboard pegs, returning to the starting point. Noam, on the other hand, divided the pegs into groups of 10 each ((A, B, C, D, E, F, G, H, I, J), (K, L, M, N, O, A, B, C, D, E), (F, G, H, I, J, K, L, M, N, O)), and this partition needs two rounds in order to get back to the starting point. Between two successive groups of 10 pegs there is an overlap, which is the common divisor (5 pegs in this case). Because Jacob's interaction with Noam was mainly through what he saw on the computer screen and how he interpreted that, Noam's example of the triangle was problematic for him because it resulted in a simple polygon, rather than a star (Lavy & Leron, 2004).

The investigation process of the case of a star with less than *n* vertices was continued into the fifth session, because the fourth session ended with the previous episode in which there was no agreement between the two students. When Noam realized that the previous example was not helpful for expressing his idea to Jacob, during the fifth session he produced six more examples that resulted in a star with less than *n* vertices; 8-star [*repeat 24 (jump 9)*]; 12-star [*repeat 24 (jump 10)*]; 5-star [*repeat 20 (jump 8)*]; 7-star [*repeat 14 (jump 8)*]; and 9-star [*repeat 18 (jump 4)*]. The need to bring so many more examples served two purposes: First, it helped clarify his idea of the common denominators to Jacob, and second, it helped in making it clear to himself.

The following excerpt is taken from the beginning of the fifth session, in which Noam gave more examples relating to the case where, for a certain values of *n* and *k*, a star with less than *n* vertices can be reached. In fact, as noted earlier, the regularity that Noam and Jacob were trying to uncover was that the number of vertices of the resulting shape equals $n / gcd(n,k)$.

[5.1] Noam: [pointing to a blank 24-peg board on the screen] you can see by the divisor, you see? Look here for a minute, OK? Look, I tell you that in the one of nine it will come out a star with 8 vertices.

[5.2] Jacob: [pointing at the same blank board] of nine ...

[5.3] Noam: No, not in a 9-board, but a jump of 9, it will come out 8 vertices ... [at last executing it on the computer] repeat 24 ...

[5.4] Jacob: Why? Because it is, like, the nearest to a divisor of 24?

[5.5] Noam: [with frustration] No only if you know [re-types the command] it will come out an 8-star because ... [counts aloud 8 vertices on the screen] look it comes out 9 [meaning: a jump of 9 will result in an 8-star].

[5.6] Jacob: Why?

[5.7] Noam: Because actually 9 ... what is its common divisor with 24?

[5.8] Jacob: 3.

[5.9] Noam: That's right.

[5.10] Jacob: What else did we have, 10? 15?

[5.11] Noam: Let's do 10 [meaning jump 10, still on a 24-board].

[5.15] Noam: Let's do 10. it will come out 12 vertices [types in the instruction repeat 24 (jump 10) and, as predicted, a 12-vertex star appears on the screen. He confirms by counting the vertices one by one, pointing at them on the screen.]

[5.16] Jacob: Because you did 10, you did 10.

[5.17] Noam: Like 2 times 12 is 24, got it? Let's try another one, OK? [changes to a 20-peg board]

[5.18] Jacob: Do a smaller one. She [the teacher] said not to raise the number.

[5.27] Noam: We need one with many divisors, you see why.

[5.41] Noam: [exploring repeat 20 (jump 8)] It will do a star with 5 vertices [on the screen appears a 5-star]

[5.42] Jacob: OK.

[5.43] Noam: No, with 5, with 5.

[5.44] Jacob: Right.

[5.45] Noam: Ah, because 8, its common divisor with 20 is 4 and the common denominator is 5 [by this last part he apparently means that the "common denominator" of 4 and 5 is 20].

[5.46] Jacob: You do 8 times 5 and 20 times 5, its 8 times ...

[5.47] Noam: You can pass it like this.

[5.55] Noam: Ah, ok, ok, let's do 14 [changes to 14 peg-board, still thinking of a jump of 8] it's got a 7 star.

[5.56] Jacob: Why? Because you divide by 2?

[5.57] Noam: No, because 2 ... because 8 is 2 times 4, and 7 is ... what you need to multiply to get 14 is 2, right? [Noam and Jacob are looking at the 14 board on the screen].

The previous excerpt demonstrates a process of successive refinement of Noam's attempts to explain his ideas to Jacob, although all along his utterances remained vague and incomplete.

At this stage of the inquiry Noam does not have a precise expression for describing the case in which a star of less than n vertices is reached. The first example Noam and Jacob tested in the excerpt is *repeat 24 [jump 9]*. Noam's utterances lead to the assumption that he thought of the pair of relationships $8 \times 3 = 24$, $3 \times 3 = 9$, which for him would mean that the command will produce a star with 8 vertices. This is evidenced first by his ability to predict the number of vertices in *all* the examples they test, and second, by his utterance [5.57].

The successive refinement of Noam's explanations can be distinguished by four significant utterances, starting with [5.5], in which he tried to explain to Jacob why *repeat 24 [jump 9]* will result in an 8-star. At this initial stage, his explanation does not refer at all to numbers' attributes and is consisted only from the assertion; "only if you know." His verification of this case is done by actually counting the vertices on the screen. The second stage is initiated in [5.17] in which he predicted that *repeat 24 [jump 10]* will result in a 12-star. At this point his explanation—"like 2 times 12 is 24"—already contains the relation $2 * 12 = 24$, but not the second relevant relation $2 * 5 = 10$ (as explained in the previous paragraph).

The third stage starts with utterance [5.45], in which they tested the command *repeat 20 [jump 8]*. Here can be seen, for the first time in the discourse, both *common divisor* and *common denominator* concepts. In fact, this is the first time that both relationships, $4 * 5 = 20$ and $4 * 2 = 8$, are mentioned, although the expression of the second is rather unclear. This utterance shows that Noam now replaced the term *common divisor* with *common denominator*. The fourth stage starts with utterance [5.57], which concerns the example of *repeat 14 [jump 8]*, resulting in a 7-star. In the latter example, Noam finally gave a relatively clear explanation of how to calculate the number of vertices of the star. Namely, that $7 = 14 / gcd(8,14)$.

From the whole process described earlier, it can be see that although Noam had difficulties in formulating his ideas, he was aided by the computerized environment to complete his sentences, and thus successfully predicted the resulting star *before* the execution of the instruction *repeat n [jump k]*. This substantiates Zazkis and Liljedahl's claim (chap. 5) that before students are able to identify the rules or algebraic relationships that underlie the structure of a pattern, they are able to use the rhythm of a pattern to learn, to understand, and to create.

Jacob, on the other hand, in trying to follow Noam's line of investigation, contributed to the process by forcing Noam to articulate his vague utterances more clearly. Jacob understood that the reason for getting a star with less than *n* vertices lies in the relation between the numbers *n* and *k*, but until the end of the fifth session, his utterances relied on the fact that he could not accomplish the desirable pattern. The most detailed explanation given by Noam is when he explained to Jacob why the instruction *repeat 14 [jump 8]* results in a 7-star [5.57].

During the analysis of the students' discourse, I kept asking myself why Noam used the term of *common denominator* when in fact, conceptually, he looked for the *common divisor*. One possible reason is the mix up with the meaning of this term in the spoken language. When one says in Hebrew that he looks for the common denominators of two people, he meant to say the *common attributes* they have that, in this case, is associated with the mathematical concept of *common divisor*.

Despite the fact that Noam uncovered the pattern regarding the case of a star with less than *n* vertices, throughout the entire discourse he chose not make any attempt to explain *why* the number of the vertices is $n / gcd(n,k)$. In this case, as in the case of the n-star, the possible reason for not trying to explain the uncovered pattern is that the students did not see the group of numbers each peg represents (Fig. 9.5). When *n* and *k* has a nontrivial common divisor then the least common multiple of *n* and *k* [$LCM(n,k)$] is less than $n \cdot k$. When the $LCM(n,k)$ is a member of the group of numbers that the first peg represents, then the jump will bring us back to the first peg, thus closing the star. This explains the connection between the numbers' attributes (having a nontrivial common divisor) and the resulting star with less than *n* vertices.

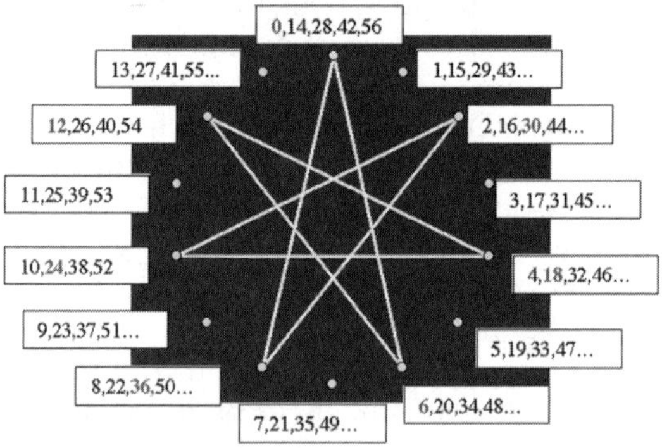

FIG. 9.5. The group of numbers each peg represents on a 14-peg geoboard—the case of *repeat 14 [jump 6]*.

At this point, from a pedagogical perspective, there is a need for the teacher to summarize the students' findings and by means of classroom discussion, to complete the formal proof of the uncovered regularity.

CONCLUDING REMARKS

Bearing in mind that collaboration among learners is not always successful, when I first looked at the discourse during the investigation process of Noam and Jacob, I had the impression that the teamwork between Noam and Jacob had failed. However, depth analysis of the teamwork shows that their collaboration was fruitful and in fact, despite (or perhaps thanks to) the differences between the students, they succeeded in uncovering mathematical regularities that are closely related to number theory concepts. This process was made possible by two main factors:

1. The students were aware both of their own limitations and their partners' abilities that could help them overcome these limitations. For example, Noam, who was aware of his limitation in verbalizing his ideas, used Jacob's utterances, corrected them if necessary, to express his own ideas. Jacob, who was aware of the rapidity and efficacy with which Noam handles the keyboard, let him do all the typing and concentrated on the inquiry process.
2. The mediation of the computerized setting of the geoboard. The computerized environment provided them with another language to communicate with. Whereas during the first few sessions of the investigation process, most of the utterances included terms of the geoboard like: "in 24-peg geoboard a jump of 3 pegs results in an 8 -side polygon," from the fourth session onward, elementary number theory concepts began to play a major part in the students' discourse. In addition, the immediate screen images that were produced, resulting from the command *repeat n [jump k]* for certain inputs of n and k, encouraged them to keep on looking for mathematical patterns.

Research shows that preservice teachers have difficulties in understanding various elementary number theory concepts (Campbell & Zazkis, 2002). Because these concepts are mainly taught in elementary school, it is reasonable to infer that elementary school students do not get proper and comprehensive instruction in them. Hence, it is important to encourage middle school students to engage in activities that deepen their understanding of number theory concepts. If these activities included means of helping students connect different mathematical topics, as we see has been the case with the geoboard, they would benefit more.

In the investigation described earlier, two students, Noam and Jacob, uncovered two mathematical regularities. The first refers to the connection between the two inputs (n and k) in the command *repeat n [jump k]* and the corresponding concept of an n-star. Because these students based their conclusions on a limited number of examples that they tested in a computerized geoboard setting, they did not succeed in actually formulating the final and complete version of that concept. That is, they did not find that the two inputs in the previous command have to be relatively prime in order to get an n-star. Although they succeeded in generalizing from examples in both cases—the n-star and a star with less than n vertices—they did not give a detailed explanation *why*.

In order to bring the students to get a complete view of regularities they uncover during their investigations, classroom discussions should follow. In these discussions, which should be monitored by the class teacher, formal proofs can be constructed in a process of negotiations between the students and the teacher. In this manner, students could gain a clearer view of the concepts involved, and could also be helped to discover more advanced concepts and relations. In the process of understanding *why* their uncovered regularities are valid, students will gain opportunities to deepen their understanding of the elementary number theory concepts involved.

REFERENCES

Ball, D. (1990). Prospective elementary and secondary teachers' understanding of division. *Journal for Research in Mathematics Education, 21,* 132–144.

Brown, G., & Yule, G. (1983). *Discourse analysis*. Cambridge University Press.

Campbell, S. R., & Zazkis, R. (2002). *Learning and teaching number theory: Research in cognition and instruction*. Westport, CT: Ablex.

Cazden, C. B. (1988). *Classroom discourse: The language of teaching and learning.* Portsmouth, NH: Heinemann.

diSessa, A. (2000). *Changing minds: Computers, learning, and literacy*. Cambridge, MA: MIT Press.

Goetz, J. P., & LeCompte, M. D. (1984). *Ethnography and qualitative design in educational research.* New York: Academic Press.

Graeber, A., Tirosh, D., & Glover, R. (1989). Pre-service teachers misconceptions in solving verbal problems in multiplication and division. *Journal for Research in Mathematics Education, 20,* 95–102.

Greer, B. (1992). Multiplication and division as models of situations. In D. A. Grouws (Ed.), *Handbook of research on mathematics teaching and learning* (pp. 276–295). New York: Macmillan.

Harre, R., & Gillet, G. (1994). *The discursive mind.* Thousands Oaks, CA: SAGE.

Lavy, I. (1999). *Understanding basic concepts in elementary number theory: Explorations in an interactive computerized environment.* Unpublished doctoral dissertation, Israel Institute of Technology, Haifa, Israel.

Lavy, I., & Leron, U. (2004). The emergence of mathematical collaboration: The role of an interactive computerized environment. *International Journal of Computers for Mathematics Learning, 9*(1), 1–23.

Leron, U. (1985). A direct approach to indirect proofs. *Educational Studies in Mathematics, 16,* 321–325.

Lester, F. K., & Mau, S. M. (1993). Teaching mathematics via problem solving: A course for prospective elementary teachers. *For the Learning of Mathematics, 13*(2), 8–11.

Martin, W. G., & Harel, G. (1989). Proof frames of preservice elementary teachers. *Journal for the Research in Mathematics Education, 20,* 41–51.

Mason, J., Burton, L., & Stacey, K. (1982). *Thinking mathematically* (rev. ed.). London: Addison-Wesley.

Movshovitz-Hadar, N., & Hadass, R. (1991). Preservice education of mathematics teachers using paradoxes. *Educational Studies in Mathematics, 21*(3), 265–287.

Noss, R., & Hoyles, C. (1996). *Windows on mathematical meanings: Learning cultures and computers*. Dordrecht, Netherlands: Kluwer.

Olson, J. (1991). Function recognition and computerized graphing. *Physics Teachers, 29*(5), 293–295.

Papert, S. (1980). *The connected family: Bridging the digital generation gap*. Atlanta, GA: Longstreet.

Sfard, A. (2000). Steering (dis)course between metaphor and rigor: Using focal analysis to investigate the emergence of mathematical objects. *Journal for Research in Mathematics Education, 31*(3), 296–327.

Sfard, A., & Kieran, C. (2001). Cognition as communication: Rethinking learning-by-talking through multi-faceted analysis of students' mathematical interactions. *Mind, Culture, and Activity, 8*(1), 42–76.

Sinclair, N., Zazkis, R., & Liljedahl, P. (2003). Number worlds: Visual and experimental access to elementary number theory concepts. *International Journal of Computers for Mathematical Learning, 8*(3), 235–263.

Zazkis, R. (1995). Fuzzy thinking in non-fuzzy situations: Understanding students' perspective. *For the Learning of Mathematics, 15*(3), 39–41.

Zazkis, R. (1999). Odds and ends of odds and evens: An inquiry into students' understanding of even and odd numbers. *Educational Studies in Mathematics, 36*(1), 73–89.

Zazkis, R., & Campbell, S. (1996a). Prime decomposition: Understanding uniqueness. *Journal of Mathematical Behavior, 15*(2), 207–218.

Zazkis, R., & Campbell, S. (1996b). Divisibility and multiplicative structure of natural numbers: Preservice teachers' understanding. *Journal for Research in Mathematics Education, 27*(5), 540–563.

Chapter 10

Overlooking Number Patterns in Algorithmic Problem Solving

David Ginat
Tel-Aviv University

Consider the following algorithmic task: Design an algorithm (computer program) for calculating whether a list of N positive integers contains two integers where one is at least twice the other. This task is considered basic for CS1 (Foundations of Computer Science 1) students. One would like to expect that students base their solution on the following mathematical pattern: The largest ratio between two integers in the list is the-max-integer/the-min-integer. The elegant algorithmic solution of the task involves the calculation of these two extremal values and their corresponding ratio. Do novice CS1 students offer this solution? Not quite.

In our experience, a considerable amount of students offer the very inefficient and cumbersome solution of comparing *every* two integers in the list, and finding the largest ratio. This solution requires an order of N^2 comparisons, rather than the order-N comparisons for calculating the maximum and the minimum in the extremal-based solution. The students seem to overlook the task's underlying number, or mathematical pattern of the largest ratio of max/min. Does this phenomenon occur in additional tasks? What are its primary characteristics? And what is the lesson for educators? The objective of this chapter is to address these questions and shed light on this phenomenon.

The displayed task is typical in computer science. It involves a computation over a sequence of elements. The required computation is of a searching kind. Searching tasks, as well as enumeration, counting, optimization, and numerical calculations are fundamental in computer science at all levels. One is expected to analyze the task's characteristics prior to the actual solution design, and base the solution on recognized mathematical patterns (Dijkstra, 1976; Gries, 1981; Oprea, 1988). A solution based on mathematical insight yields correctness and efficiency. However, mathematical insight is not always immediate. Due to lack of awareness of its fundamental role, one may "skip it," and hastily offer a less suitable, or even an unsuitable solution.

In the course of our work with computer science students at various high school and college levels, we noticed a repeated lack of awareness of the need for solid mathematical insight. Students demonstrated time and again a lack of quest for essential number and sequence patterns, such as the extremal pattern displayed earlier. They often offered noninsightful, naïve solutions to basic algorithmic tasks, and erroneous solutions to more involved tasks. The reasons for the students' inappropriate behaviors vary. In this chapter, we demonstrate and illuminate several facets of their behaviors.

We display examples of student solutions for seven different algorithmic tasks, which involve the computation kinds mentioned previously. Some of the tasks are basic, and others are more involved, including two that require game strategy design. They are all based on fundamental notions of discrete mathematics, such as modulo arithmetics, binary representation, primality considerations, and combinatorial characteristics, which repeatedly occur in algorithmic problem solving. The suitable solutions of the tasks require mathematical insight in the form of simple number and sequence patterns. Yet many students overlooked these patterns, and offered inefficient or even erroneous solutions, which they argued correct. We divide the task–solution presentation into three primary categories, which capture different characteristics of the students' inappropriate behaviors. We then analyze the students' solutions in light of Schoenfeld's model of problem solving (1985, 1992), while explicitly referring to the model's components of resources, heuristics, control, and beliefs.

The presented solution examples were collected in several studies, which we conducted with 11th graders, 12th graders, and college beginners, some of which were relatively competent and particularly motivated. All the students completed their CS1 course, and some completed their CS2 (Fundamentals 2) course. In the studies, we posed tasks, collected the students' written solutions, and interviewed some of them on their solutions. We describe the students' written solutions, and present their comments and parts of the interviews.

In each of the next three sections, we present students' solutions and comments on two algorithmic tasks. The title of each section corresponds to a particular characteristic of student difficulties, or tendencies in solving algorithmic tasks. Each characteristic is reflected by the students' unsuitable algorithmic solutions. We present for each task the students' unsuitable solutions and the elegant, desired solution, which derives from underlying number patterns. In the concluding two sections, we discuss the presented results in a broader perspective and offer suggestions to educators for addressing the phenomena illuminated in the earlier sections.

LACKING MODULO ARITHMETICS

Algorithmic computations involve repeated utilization of modulo arithmetics. Some simple examples already at the very beginning of programming are the calculation of the number of buses needed for carrying a given number of travelers; the calculation of the landing time of a flight, given the takeoff time and the flight duration; and the calculation of the floor number of a hotel room, given a 4-digit room number in which the first two digits indicate the floor.

These calculations require the computations of the quotient and the remainder of integer division, using the corresponding operators of *div* and *mod*. Students are introduced to these operators already in elementary school. Are they acquainted with these operators' utilization? Apparently not enough. Although the direct computations of the quotient and the remainder are elementary and easy to understand, students' ability to use the div and mod operators in diverse occasions is considerably limited. In algorithmics, one may sometimes bypass the relevant and elegant invocation of these operators with an exhaustive search through an unnecessarily large range of values. Such an approach is undesirable, yet students follow it, and avoid, or overlook number patterns that are yielded from modulo arithmetics. We show this phenomenon below.

This section displays two elementary algorithmic tasks, which require basic calculations using div and mod. Although elementary, many students approached these tasks inappropriately, employing an exhaustive search, and yielding inefficient, and possibly even impractical solutions.

Task: Day of the Week

Develop an efficient algorithm (computer program) for which the input is composed of two integers—the current day D, of the week and the duration, N in days, of a particular (possibly very long) event, and the output is the day of the week on which the event will end.

Example. For the input 3 50 (i.e., Tuesday is the current day, and the event duration is 50 days), the output will be 4 (i.e., Wednesday).

Students mostly offer three solutions, two of which are inefficient and inelegant. We call these solutions "no modulo arithmetics," "cycle truncation," and "direct remainder." The first solution involves no mathematical insight, the second is based on a relevant, yet still limited insight, and the third is based on elegant capitalization on relevant number patterns.

Solution 1: No Modulo Arithmetics

Although it seemed surprising to us, and may surprise the reader as well, many novices write a program that involves a loop, which increments a variable day N times, after initializing day to D, the starting day. Every time the variable day reaches 7, the next increment yields 1. This solution is correct, but involves no modulo arithmetics. The computation time is of order-N, which is very inefficient.

We interviewed two students about this solution:

Guy: The algorithm is very simple—increment the day variable iteratively N times, while reducing it to 1 once it reaches 7.

Interviewer: So if N is 10,000, there will be 10,000 iterations, is this really necessary?

Guy: Maybe it is not the most efficient way, but I am not sure how to compute it directly.

Interviewer: Do you think that it may be directly computed?

Guy: Yes ... but the direct way requires an arithmetic expression, which is too complex for me.

Interviewer: Why?

Guy: It involves some div and mod expression which I am not sure how to phrase.

Another student added an additional explanation:

Irit: In this task there is a repeated cycle of length 7, and after every 7 increments we get back to the initial point. But I do not know how to translate this into some kind of direct computation of the last day, and in any case, my computation is simple and correct, the computation will just advance up to the necessary halting point.

These students "managed" to solve the problem without reaching any mathematical insight. They did not capitalize on any relevant number pattern. Their solution encapsulates a linear progress, using a series of increments, which could be avoided by recognizing that the relevant component in the required computation is the remainder of dividing N by 7. The first student, Guy, referred to the notion of div and mod, without clarifying their relevance. The second student, Irit, did indicate the notion of a cycle, but overlooked the relevance of the remainder. All in all, these students "by-passed" the required pattern by an inefficient and inelegant iterative scheme, which served as a "replacement" for mathematical insight.

Some students did observe the relevance of the notions of cycle and remainder, and combined modulo arithmetics. However, they employed it only to a limited extent.

Solution 2: Cycle Truncation

A small amount of students noticed the notion of a cycle, and offered a scheme that "truncates from N all the week cycles." That is, they subtracted from N the largest multiple of 7 that is smaller or equal to N; using the expression: N – 7 · (N div 7). Then, their computation scheme iterated N – 7 · (N div 7) increments over `day`, starting at the value D. Interestingly, they did not use the mod operator. We interviewed one of these students.

Or: We may eliminate the highest multiple of 7, since advancing by this value brings us back to the initial day D … then, what matters is only the remaining number of days, which is smaller than 7.

Interviewer: Why do you iterate over this remaining number of days?

Or: We have to increment D carefully, and reduce it to 1 if it exceeds 7.

Interviewer: Do you have to iterate?

Or: Perhaps not, but I was not sure how to do it directly, so I chose a safe way.

Interviewer: What was the difficulty here?

Or: The sum: D + [N – 7 · (N div 7)] may exceed 7, and then I have to truncate it, and I am not sure how to do it with a corresponding expression.

Or did recognize the relevance of adding (N mod 7) to D. However, he did not express this observation using the mod operator. Instead, he used the expression N – 7 · (N div 7). Then, he did not recognize the relevance of a mod operation on the value of D + [N – 7 · (N div 7)]. This lacking insight yielded an unnecessary iterative computation. Thus, although he expressed some mathematical insight, this insight was insufficient, and lacked clear recognition of the role and the computation of remainders. Again, the lack of mathematical insight was "overcome" by a simple iterative computation.

Some students clearly recognized the role of remainders, formulated the corresponding arithmetic expression, and avoided any iteration.

Solution 3: Direct Remainder

The elegant way of performing the relevant calculation is by directly calculating the remainder of dividing D + N by 7. This is expressed in the following pattern.

Remainder Pattern. The desired output is derived from the expression (D + N) mod 7. If the value of this expression is 0, then the output should be 7 (i.e., Saturday); otherwise, the output should be exactly the expression's value. (It is possible to avoid the "if" addition, by slightly changing the above expression to: [(D + N – 1) mod 7] + 1. Yet, we consider the solution that includes the "if" addition satisfying enough.)

The key point in this task was to recognize the remainder pattern, and employ rather basic modulo arithmetics. The computation yielded from this arithmetic requires no iterative scheme, but rather a simple mathematical expression that directly yields the desired result.

The next task also requires simple calculations with modulo arithmetics, and again, one may avoid it by looping. One may sometimes loop over a whole range of values, while avoiding the relevant mod operation.

> *Task: Sequence of Multiples*
>
> Develop an *efficient* algorithm (computer program) for which the input is composed of two positive integers, x and y, where x < y, and the output is the sequence of decreasing positive multiples of x that are smaller or equal to y.
>
> *Example.* For the input 100 597, the output will be 500 400 300 200 100.

As in the previous task, students offer three solutions, two of which are inefficient and inelegant. We call these solutions "exhaustive range search," "remainder range search," and "concise generation." Again, the first solution involves no mathematical insight, the second is based on a relevant, yet still limited insight, and the third is based on elegant capitalization on the relevant number patterns.

Solution 1: Exhaustive Range Search

Students identify that the relevant range for the output values is the range [x,y]—a few actually extend this range, unnecessarily to [1,y]. Their solution is very simple; "loop over the range elements from y down to x, and check for each element if it is a multiple of x (if it is—then output it)." This solution is correct, but considerably inefficient. The number of iterations is y – x + 1, and may be significantly larger than the output size, which is y div x (e.g., the case of x = 1,000, y = 30,000). The following is a typical student explanation.

Yigal: My solution goes over each of the values from y to x, and outputs the relevant ones.
Interviewer: Why go over each of these values? Many of them may be irrelevant, perhaps it is possible to directly generate only the relevant ones?
Yigal: Well ... perhaps, but this requires finding the first relevant one, and then orderly finding all the others.
Interviewer: So, why not do that?
Yigal: I do not know how to directly find the first one, and then, even after finding it, I need to make sure that I find all the others and do not miss any of them. So the loop in my solution guarantees that I do not miss any multiple of x.

The terminology used by Yigal is that of "finding" elements, and not generating them. It seems natural to him that an exhaustive search over the relevant range of values will yield an ordered "finding" of the necessary output elements. His explanation reflects a lack of ability to directly calculate both the first output element and the elements that follow. The calculation of the first element requires recognition of a remainder pattern. The calculation of the rest of the elements is much simpler and only requires recognition of the constant (x) difference between every two consecutive output elements. Yigal overlooked both of these patterns.

The students in the next solution category were still unable to recognize the former pattern, but did recognize the latter one.

Solution 2: Remainder Range Search

Many students recognized the constant difference between consecutive output elements. Yet they were unable to directly calculate the first element. In order to overcome this difficulty, they offered a loop that starts at the value y and repeatedly decreases it by 1, "until a multiple of x is found." This multiple is the first output element. From that point on, their scheme directly generates the rest of the output elements.

This solution is more efficient than the previous one, but still involves an unnecessary search. This search is shorter than the exhaustive search in the previous solution. Yet, it reflects again the students' way of "overcoming" the lack of mathematical insight by a systematic search. In interviews about this solution, students indicated that they felt that there may be a direct way of calculating the first output element, but could not formulate a corresponding arithmetic expression. Therefore, they designed the previous searching loop. This loop "compensated," in a sense, for the lack of recognition of the remainder pattern.

Some students did formulate the relevant arithmetic expression, using rather simple modulo arithmetics.

Solution 3: Concise Generation

The elegant way of generating the output elements is based on the following two number patterns.

- *Remainder Pattern.* The largest multiple of x, which is smaller or equal to y is at distance y mod x from y. The value of this multiple of x may be calculated in two ways: y – (y mod x); or x · (y div x).
- *Difference Pattern.* The difference between two consecutive multiples of x is x.

The elegant computation derived from these patterns involves a loop in which all the output elements are orderly generated. The number of iterations is y div x, which is considerably smaller than the number of iterations of the previous searching loops. This was enabled by capitalizing on the above simple number patterns, which were expressed by modulo arithmetics.

IMMEDIATE DESIGN PATTERNS

The basic building blocks of algorithmic solutions are design patterns (Astrachan, Berry, Cox, & Mitchener, 1998; Linn & Clancy, 1999). Design patterns are recurring computational schemes, such as searching, counting, and accumulation. They may appear in simple or complex forms, depending on the task at hand and the required computation. They are the actual operational constructs that compose an algorithm or computer program. On reaching insight into an algorithmic task, a problem solver designs an algorithmic solution by invoking and composing suitable design patterns.

However, insight levels may vary. One may only reach preliminary insight and already have a design idea, involving relevant design patterns. Alternatively, one may obtain further insight and only then approach the design stage. The design patterns in both cases may be similar or different. In the former case, they will usually be directly "extracted" from the task description. In the latter case, they may be based on hidden unfolded mathematical patterns (Ginat, 2001b).

The two very different solutions of the day of the week task involve different design patterns. The naïve, noninsightful solution encloses a "pair-generator" design pattern, whereas the insightful solution encloses the utilization of a min/max design pattern. The relevance of the former pattern may be directly extracted from the task description, which states the requirement of finding out whether among all the possible integer pairs there is a particular desired one. The relevance of the latter pattern is seen only after reaching suitable mathematical insight into the task.

The second task in the previous section—sequence of multiples—also involved different design patterns. The elegant solution was based on a "generator" design pattern. The inelegant and less efficient solutions were based on the design pattern of "linear search." Thus, a "generator" design pattern was less suitable in the day of the week task, but most relevant in the sequence of multiples task. The design pattern of "linear search" was less suitable in the sequence of multiples task, but may be very relevant in other tasks. Mathematical insight is a primary factor in the selection of the design patterns on which an algorithmic solution is based.

This section displays two rather basic algorithmic tasks, which involve counting, searching, and accumulation design patterns. We start with a basic counting task, of counting the number of substrings in a given string. The students were *explicitly* requested to design an efficient solution, both in terms of computation time and memory usage (i.e., time and space complexities; Cormen, Leiserson, & Rivest, 1991).

> *C-C-C Substrings*
>
> Develop an *efficient* algorithm (computer program) for which the input is a long string of characters, and the output is the number of strings that start with a *c*, end with a *c*, and contain exactly one more *c* between the two ends.
>
> *Example*. For the input *abcbcacac*, the output will be 2.

Students mostly offer two solutions, which we call *whole-element counting and prefix counting*. The first is a naïve noninsightful solution and the second is the insightful one.

Solution 1: Whole-Element Counting

Many novices approach the task by immediately combining straightforward searching and counting design patterns. Their solution is based on searching for the next whole substring, that is, the next "trio" of *c...c...c*. Their algorithm keeps the whole input string in memory and performs repeated "runs" over this string. Each identified sub-string increases the substrings counter by 1.

When interviewed about the above solution, Adi offered the following explanation:

Adi: My idea is simple—collect the substrings one-by-one, by recognizing each one of them separately. This processing scheme is called for by the task description.
Interviewer: Is this an efficient way to solve the task?
Adi: I think so.
Interviewer: Perhaps there is a better way, without explicit collection of whole substrings?
Adi: I do not think so, as the counting process must involve a separate increment of the counter for each valid substring.

Another student added the following comment:

Amitai: This solution is natural and becomes obvious as you read the task description; since the computation process must involve a separate recognition of each valid substring, with an initial *c*, and intermediate *c*, and an ending *c*.

The repeated-runs solution may indeed seem natural, or even "called-for," by those who do not seek any insight into the given task. Their design appears to be directly extracted from the task specification—each of the defined elements (substrings) should be found and counted. A simple composition of the design patterns of searching and counting yields the corresponding computational scheme.

It looks as if the task phrasing was directly "translated" into a corresponding algorithmic scheme that involves the relevant design patterns. This "translation phenomenon" is similar to some extent to related phenomena reported by mathematics educators. Briars and Larkin (1984) as well as Hegarty, Mayer, and Monk (1995) displayed problem solving strategies of less competent students, which were based on identifying keywords and directly translating phrases in the description of word problems. Briars and Larkin named it "the keyword method," and Hegarty et al.—"the direct translation strategy."

The immediate algorithmic solution displayed above reflects a similar behavior—an explicitly defined element should be counted, therefore the task "translation" into a solution implies a simple composition of the design patterns of searching (for an element) and counting. However, if one seeks mathematical insight prior to the design, one may notice a combinatorial pattern, which significantly improves, and simplifies the previous solution. The students who offered the above solution overlooked this pattern; yet other students did recognize it and offered the better solution.

Solution 2: Prefix Counting

In examining several examples, one may notice a simple relation between the number of valid substrings and the numbers of *c*'s in the input string, as stated in the following mathematical pattern.

> *Prefix Pattern*. Each *c* in the input, apart from the last two *c*'s "opens" (i.e., is the prefix of) *exactly* one substring of the form *c...c...c*. Thus, the total number of sought-after substrings is two less than the total number of *c*'s.

This pattern is not exactly a number pattern, but rather a sequence pattern. However, it encapsulates the very relevant combinatorial pattern of "the number of different subsequences of K consecutive elements in a sequence of length N is N – K + 1." In our case, K = 3.

The algorithm design that follows from the previous pattern encloses simple searching and counting patterns; but the searching is not for the substrings defined in the task phrasing, but for single *c*'s. Thus, this insightful solution involves a single "run" over the input string, and therefore does not require keeping the whole string in memory. It is simpler and much more efficient than the immediate earlier one, both time-wise and space-wise.

The second task of this section involves a question of the form "can a given element be composed of certain components?" One way of answering this question is by systematically generating combinations of the relevant components, and checking whether some combination yields the given element. Another way is by trying to decompose the given element into components of the desired kind. These two very different methods encapsulate opposite ways of reasoning. The former may be more relevant for those who obtain only limited mathematical insight. The latter may be relevant for those who gain insight and recognize a suitable decomposition way.

> *Task: Powers of 2*
>
> Develop an *efficient* algorithm (computer program) for which the input is a positive integer N, and the output is a message indicating whether the input integer may be the sum of three *different* powers of 2.
>
> *Example*. For the input 10 the output will be "No," and for the input 21 the output will be "Yes" (as 21 = 1 + 4 + 16).

Students offer two solutions, which we call *exhaustive search* and *binary representation*. The first is based on searching for some combination of powers of 2 that yields N. The second is based on the opposite reasoning direction, of decomposing N into relevant components.

Solution-1: Exhaustive Search

A common solution among many novices was based on an ordered generation of triples of powers of 2, starting with the triple <1,1,1> and advancing systematically up to a triple of powers of 2 that is closest to N. The components of each triple are summed and compared to N. Upon equality, they are checked to be different from one another. This exhaustive search, over all possible triples is often written using three nested loops (one for each component in the triple). This approach may indeed yield the correct output, but it is inefficient and inelegant. In addition, many students demonstrated difficulties in such generation of triples, and offered erroneous computer programs. The following interview demonstrates the rationale of students who offered this approach.

Ofer: My scheme generates all the possible combinations of powers of 2. If there is a valid combination that yields N, then my scheme finds it. I know that this scheme is not optimal since it also generates triples of equal powers of 2, but the scheme recognizes such triples and does not output them. I chose not to generate only different powers of 2 in each triple, since I was not sure about my code's correctness in this case.

Interviewer: Would a code that generates only triples of different values be optimal?

Ofer: I think so, since it will search only through valid combinations of powers of 2.

Interviewer: The scheme you indicate is based on some search; is it necessary to search?

Ofer: I think so, the task asks to calculate whether there is a valid combination of powers of 2, so I have to go through the possible combinations and check if there is a valid one.

Ofer, and many others did not consider another computation way other than a systematic search. It seemed natural to him that this was the way to go, because "the task asks to calculate whether ...." His focus on efficiency is only related to the avoidance of generating invalid elements for the search process. It seems that not only did he not seek a different solution way, he did not think that there may be one. As with the previous example, here too students seemed to be caught up with some phrasing in the task description, which led them directly to a particular solution scheme. In this task, it is a systematic generation of possible combinations, using a rather involved "generator" design pattern, for yielding the triples for the exhaustive search.

However, there is another more efficient and more elegant way of solving the task. It is based on basic properties of powers of 2, and the binary representation of integers. Some students recognized and followed this way.

Solution 2: Binary Representation

The representation of information in a computer is based on binary codes. Each integer has a unique binary representation. Students learn and practice conversions from decimal to binary representation, and vice versa, already in the very beginning of CS1. In particular, they learn that each bit in the binary representation of an integer indicates whether the bit's corresponding power of 2 is present in the sum of powers of 2 that yields that integer (1 indicates "present," and 0 indicates "absent"). This characteristic yields the following number pattern.

Binary-Representation Pattern. Because the binary representation of an integer N is unique, N is the sum of three different powers of 2 if and only if there are exactly 3 bits of value 1 in the binary representation of N.

The input integer N may be transformed into its binary representation, and the number of 1 bits will be counted. This implies a simple "decomposition" design pattern, of decomposing an integer into its composing powers of 2—first the largest one, then the second one, and so forth. If the decomposition yields exactly three powers of 2, then the answer should be "yes"; otherwise—"no."

Unfortunately, many students did not relate their knowledge of binary representation to the above pattern. They overlooked this pattern, and offered the previous less suitable solution.

HASTY GENERALIZATION

One of the more useful heuristics in solving problems is generalization from examples (Polya, 1957; Schoenfeld, 1985). In algorithmic problem solving, problem solvers often examine various examples of input–output relations during the primary task-analysis stage. Indeed, such examination may yield valuable insight, which may include the recognition of general, underlying patterns. Although such generalization may be very fruitful, one has to be very careful, and beware of premature convictions. Recognized observations should be verified and justified, based on solid mathematical patterns. However, in our experience, student awareness of the essential need for solid mathematical justification is limited. As a result, premature, unfounded, and even hasty convictions are apparent. In this section, we present two examples of this inappropriate occurrence.

Our experience with this occurrence involves students at all levels, including competent ones. We chose to display here our experience with these students. Thus, the two examples presented in this section include more challenging tasks than those presented earlier. They involve mathematical games, for which students had to devise winning strategies. We posed these games to relatively competent students, some of whom participated in an evaluation activity toward the international youth Olympiad in Computer Science. Although relatively competent, some of these students lacked sufficient scientific discipline.

The first game involves the notions of coprime integers. Games arouse motivation and intuition. In algorithmic problem solving, games may serve as colorful, attractive tools for illuminating core concepts. One such concept is that of loop invariants (Ginat, 2001a). This concept will apparently be most relevant here.

> *Coprimes Game*
>
> Given a line of N (≥ 15) consecutive integers {k, k + 1, k + 2 ...}, k > 1, two players remove the integers—one at a time—in alternating turns, until only two integers remain. If the two remaining integers are coprimes then the first player (who starts the game) wins; otherwise—the second player wins. Devise a computer program, which first inputs k and N from a user, who is the opponent, and decides whether to be the first or the second player. Then, the program plays against the user such that it always wins.
>
> *Example of a mini-game*. For the initial input 3 5 by the user, which implies that the game is played on the integer line; 3 4 5 6 7, your program may choose to be the first player, and remove the 6; the user (opponent) may remove the 5, then the program may remove the 4, leave the two coprimes 3 and 7, and win.

In an initial analysis of the game, one may notice that if N, the initial number of integers is odd, then the first player has an extra move. This may be a significant advantage. Thus, many problem solvers are led to seek a winning strategy for the first player, in the case of odd N. The case of even N is subtler. We first display two student solutions for the case of odd N, and call them *number of divisors*, and *coprimes adjacency*. Then, we describe students' unfruitful attempts for the case of even N, and display the pattern underlying the elegant solution.

Solution 1: Number of Divisors

One approach that many students follow is that of removing integers according to their number of divisors. Their goal is to leave at the end of the game two integers, which have "as low as possible number of divisors (and hopefully one of these integers may even be a prime)." This process, they claim "yields coprimality." Thus, in its first turn, the first player (their program) removes the integer with the largest number of divisors, in its next turn, the first player removes the integer with the second largest number of divisors, and so forth. Toward the end of the game, there may be a few alternatives of integers with an equal number of divisors. In this case, their program carefully selects the better alternative, based on the integers removed by the opponent. Many students tried this strategy on examples.

Amir: I examined several example games, and noticed that the number of divisors seems like a most relevant parameter. This yielded my "divisor-based strategy." I then checked it first on a small scale, on the initial integer list 2 3 ... 9 10; then on a larger scale, on the initial list: 10 11 ... 23 24. In these examples, and others, my algorithm works.

Interviewer: So, does this mean that it is correct, and will work for any initial list?

Amir: Well ... I have no formal proof. But you see that it makes sense, it leaves two disjoint integers in the end of the game.

Interviewer: Did you try looking for a falsifying example?

Amir: Not really, but I examined quite a few examples, and this strategy always worked.

The interview reflects a common students' approach, according to our experience. Students examine several examples, "identify" a recurring characteristic, and generalize accordingly. Quite a few, like Amir, do not seek falsifying evidence, but rather solely focus on supporting evidence. This occurrence is typical of problem solvers who rather quickly reach an unfounded intuitive conviction (Fischbein, 1987, 1999).

We showed Amir, and other students who offered his solution, the input example 44 19; that is, a game starting with the initial list 44 45 ... 61 62. This example falsifies his strategy. The first player removes the numbers 60, 48, 56, ... 45 ..., while the second player keeps removing odd numbers, but not 45 (he lets the first player remove this number). Eventually, before the last turn by the first player, the remaining numbers will be 46, 58, and 62. They are all even, and therefore the first player loses.

Amir was very surprised. Yet, his response was a local "patching" of his initial algorithm, so that it "will deal with game scenarios that involve odd integers with many divisors like 45...." He still was unable to prove the correctness of his algorithm, but this did not bother him much. Thus, on seeing some falsifying evidence, he was reluctant to reexamine his original solution, but rather focused on some local refinement. This kind of behavior is similar to behaviors seen in previous mathematics education studies, which demonstrated and described students' one track mind behaviors (Lithner, 2000; Schoenfeld, 1992). The correct solution for this task requires more insightful mathematical examination, which will yield a solid number pattern related to coprime properties. The students who offered the previous approach overlooked such a pattern.

Solution 2: Coprimes Adjacency

One relevant characteristic of winning strategies in mathematical games is that of keeping a property invariant throughout the game (Ginat, 2001a). For example, the winning strategy in the game of Nim is based on keeping a symmetrical state after each one of the winner's moves. Thus, it may be beneficial here to seek a relevant invariant pattern on which to base the winning strategy.

In our game, a relevant invariant pattern may be based on "coupling" each game integer, except for one, with a *unique* coprime "buddy." One integer will not be coupled, because N is odd. The first player will remove the uncoupled integer at the beginning of the game, and her response to each of the opponent's removals will be the removal of the "buddy" of the opponent's removed integer. This requires a simple coprime "buddies" scheme. Such a scheme may be based on the following very basic number pattern.

Adjacent Coprimes Pattern. Every two successive integers are coprimes. Thus, in the case of odd N, the first player may remove the smallest integer at the beginning of the game, and then keep the following invariant property throughout the game: After every move by the first player, the remaining integers may be partitioned into pairs of successive coprimes, of the form $<i,i+1>$. The invariant yields the winning strategy. It also encapsulates the justification of the strategy's correctness, and elegantly demonstrates how design and proof may be done hand-in-hand in algorithmic problem solving. One who recognizes the relevance of the pattern and invariant offers a correct, simple, and very elegant winning strategy.

In attempting the case of even N, students noticed that exactly half of the numbers in the beginning of the game are even. This led many to seek a winning strategy for the second player. They offered various heuristics to try to reach two even numbers in the end of the game. One offered heuristic was that of initially removing prime numbers; other heuristics involved various ways of removing odd numbers. None of these heuristics was mathematically proved. Although their primary intuition of trying to win as the second player was relevant, there was a number pattern that was overlooked by almost all of the students.

There is no way of guaranteeing that the two remaining numbers at the end of the game will be even, because the first player may remove all the even numbers, apart from one. The key question is whether there is a way for the second player to yield two noncoprimes that are not even, if the first player chooses to "go for" the even numbers throughout the game. At this stage, one should notice that N, the length of the initial list of numbers, is at least 15. This is an important observation, which is directly related to the following number pattern.

Multiples-of-3 Pattern. There are two *odd* multiples of 3 in every 12 successive integers.

This pattern yields the key for the winning strategy of the second player. If the first player "goes for" the even numbers, then the second player will leave at the end of the game two odd multiples of 3; otherwise, the second player will leave at the end of the game two even numbers.

The second example in this section is a game of integer removals. This game demonstrates once again the essential role of invariance and number patterns. The game was posed in the 1996 International Olympiad in Informatics.

> *L–R game*
>
> Given a line of N random integers, where N is even, two players collect the integers in alternating turns. Each player, in turn, collects an integer either from the left (L) end or the right (R) end of the line, according to her choice. The game ends when all the integers are collected. You start. Your goal is that your sum, of N / 2 collected integers be no less than your opponent's sum.
>
> *Example:* For the line of integers: –1 3 9 4, you may first remove the L end (–1); your opponent—the R end (4); you—R (9); and your opponent—L (3). Your sum will be 8 and your opponent's—7. Goal obtained. If you start by removing the 4, you may not obtain your goal. If the leftmost integer was –2, instead of –1, you would obtain the goal by reaching the same sum as your opponent.

Students offered various solutions to this task. One unsuitable common solution was based on a heuristic, which we call the *delta rule*. The elegant and correct solution is based on a hidden number pattern of interleaved locations.

Solution 1: Delta Rule

Students noticed that "selecting the larger end in each move does not work." After playing the game with various examples, they offered an alternative rule—"to make an L/R choice not according to the larger L/R end, but according to the larger L/R delta." They consider the *delta* of an L/R end as the difference between the end integer and the next-to-end integer; for example, "the value that can be gained minus the value exposed." In the task description example, the initial L-end delta is –1 – 3 = –4 and the initial R-end delta is 4 – 9 = –5. Because the L-end delta is larger, the first player's move should be L.

The delta rule does indeed "work" in many examples, both simple and complex ones. Yet, does it always "work"? Can one convincingly justify its correctness? Below is an interview with Noam about his solution.

Noam: It took a me a bit to realize this rule. I examined quite a few examples, and observed that it always works, so I guess that it is right.

Interviewer: Did you try to look for, or generate, an example that may falsify your rule?

Noam: I tried various examples, but did not seek a particular falsifying example.

Interviewer: Why is this rule correct?

Noam: First, I tried it on various examples, and it worked. Then, you can see that it makes sense. In each turn, I have a choice—left or right, it may not be clever to expose a large integer, which the opponent will then collect, unless the integer I collect is also large. So, the thing to do is a gain–loss calculation on both ends.

Interviewer: You only examine the ends. Perhaps you need to look at the whole line?

Noam: I do not think so, following the argument I just presented.

Interviewer: But perhaps this local gain–loss calculation will bring after several moves the game to a point in which no matter what you do, you lose?

Noam: No. Since I have control throughout the game—in each of the moves.

Interviewer: Can you prove that you have control of the game after each move?

Noam: Well … the rule shows you how I have control.

Interviewer: Is this a proof?

Noam: Not formal, but it is convincing, it specifies the underlying motivation …

Noam looked at various examples. Indeed, his rule "worked" in many examples. However, he did not carefully seek a falsifying example. In addition, his justification is rather intuitive, and more importantly—"operational." He justified correctness by reiterating the rule and its motivation. He did not offer an "assertional" argument, which refers to the global game state after each move. This gap between undesired operational arguments and convincing assertional arguments is typical of computer science students (Dijkstra, 1989; Ginat, 2003). Noam's game strategy cannot be proved because it is erroneous. The following example falsifies Noam's strategy.

Let the initial line of integers be 10 20 59 61 23 12. The first player will collect the 10 (because 10 – 20 > 12 – 23), the second player—who may follow any strategy she wishes—may collect the 20, the first—59, the second—61, the first—23, and the second—12. The total amounts collected are 92 by the first player and 93 by the second player. The first player loses.

Noam's strategy does not capitalize on a solid mathematical pattern. He and others overlooked a very relevant number pattern, which is the key to the correct solution, as displayed in the next solution.

Solution 2: Invariant Locations

We already showed in the previous game that an effective approach for designing a founded game strategy may be based on an invariant pattern, which will be kept by the game winner. Such a pattern may force the game into very specific game scenarios, which will guarantee success. In this game, a line end is removed in every turn. Line ends are particular locations. In carefully examining the first few moves with respect to line locations, one may notice the following. After the first move, the locations of the new line ends are of the same parity—either 1 and N-1, or 2 and N (recall that N is even), after the second move (which is the first move of the second player)—the line end locations are of different parities; after the third move—again the same parities, and so forth. The latter observation is based on the following number pattern.

Interleaved Parities Pattern. The parities of successive integers are changed "back and forth." The previous observation and pattern lead to an illuminating, constructive invariant.

Locations Parity Invariant. After every *odd* number of moves, the line end locations are of the same parity; and after every *even* number of moves, the line end locations are of different parities.

The first player may elegantly capitalize on the previous number pattern. Because the first player always approaches the game line following an even number of moves, she can choose in *each* of her turns between an integer that is in an initially odd location and an integer that is in an initially even location. Therefore, she can collect throughout the game *all* the integers in the initially odd locations, or *all* the integers in the initially even locations.

Thus, the first player's strategy will be as follows: At the beginning of the game, compare the sum of the integers in the odd locations with that of the even locations. If the former is larger, then go for the odd locations throughout the game, otherwise—go for the even locations. This strategy will lead the first player to a win in the above example, by first collecting the integer 12 and later on—the integers 20 and 61.

The above elegant strategy was elicited by the locations parity invariant, which captured the core characteristic of the integer removal process. As in the previous game, a basic number pattern was the key to the solution's underlying invariant. And, the invariant yielded not only the winning strategy, but also the assertional justification of the strategy's correctness.

DISCUSSION

The previous three sections displayed a variety of student comments and solutions for six different algorithmic tasks. The student solutions varied considerably. Some solutions were incorrect, some were naïve and inefficient, and some were correct, elegant, and efficient. The desired, and elegant solutions were all based on solid mathematical insight and illuminating number patterns. The undesired solutions stemmed from inappropriate problem-solving tendencies, which yielded overlooking of mathematical patterns, and reflected a lack of practice and awareness of mathematical insight.

The inappropriate problem-solving tendencies occurred in various ways. In the first two tasks, day of a week and sequence of multiples, they involved exhaustive search rather than direct calculation, using modulo arithmetics expressions. In the third and fourth tasks, c-c-c-substrings and powers of 2, they stemmed from quick, noninsightful invocations of design patterns, which were directly associated with the task phrasings. In the last two tasks, the coprimes game and the L R game, they involved hasty generalizations, based on partial evidence of empirical examples. In what follows, we further elaborate on these inappropriate tendencies in light of Schoenfeld's problem-solving model (1985, 1992) and its components—resources, heuristics, control, and beliefs.

Resources

The resources required for solving the tasks involved elementary number patterns and basic algorithmic (programming) design patterns. The elementary number patterns were related to extremal values, remainder calculations, basic sequence combinations, binary representation, and parity and primality characteristics of integers. The basic design patterns were mostly those of linear searching, counting, and accumulation. The students were well acquainted and experienced with all these resources. Thus, familiarity with these required resources was not an obstacle.

One significant element, which we partially consider as a resource was lacking. It was the notion of mathematical patterns as a means for illuminating task characteristics, yielding suitable design, and arguing its correctness. We regard this element as both a resource and an entity that belongs to the components of control and beliefs in Schoenfeld's model. Students seemed to be neither acquainted enough with mathematical patterns specification, nor aware of their core illuminating role. This is an unfortunate difficulty among computer science students (Ginat, 2004).

The patterns in the suitable solutions of the various tasks were all specified in a concise assertional way. The assertions encapsulated basic integer patterns, which are fundamental in discrete mathematics. In the first and second tasks, they involved short modulo arithmetics expressions, with div and mod operators. In the third and fourth tasks, they involved a simple combinatorial measure of subsets, and the fundamental link between binary representation and the powers of 2. In the fifth and sixth tasks, they involved basic properties of successive integers—their coprimality and interleaved parity. All these concise integer patterns led to efficient and elegant solutions.

It was largely evident in all the students' unsuitable solutions, that many did not attempt the assertional way and did not seek these most relevant integer patterns. We believe that students' familiarity and experiences with the assertional way and mathematical patterns should be considerably enhanced. We further refer to that in the next, concluding section.

Heuristics

The heuristics required to solve the given tasks were rather essential. They involved careful examination of simple cases (which could be helpful in the c-c-c-substrings task), decomposition of elements (powers of 2 task), and careful derivation from the data (the game tasks). All these heuristics were useful for illuminating the task's hidden, underlying mathematical patterns. Students were all familiar with these heuristics. Yet, many did not invoke them. We attribute the difficulty to their control behavior, which was too often dictated by following a single, hasty solution direction, without considering alternative directions.

Control

One primary element that distinguishes novices from experts is their limited and deficient application of control (Schoenfeld, 1992). This also seemed evident in the results presented here. Two major aspects were apparent in the incompetent solutions of the six tasks—a hasty, rather narrow, one-track train of thought; and the lack of careful analysis and verification.

In the first two tasks, students offered very simple, yet inefficient iterative solutions, instead of carefully devising modulo arithmetics expressions. Instead of attempting suitable expressions, which directly yield the desired results, they chose to devise very simple loops of exhaustive search. Although this approach solved the problem, it is definitely undesirable, as it yields noninsightful, inefficient, and inelegant computational schemes, which become impractical and unacceptable at later stages of algorithmic problem solving.

In the third and fourth tasks, students immediately turned to familiar, basic design patterns, and overlooked the relevant number patterns. In a sense, the earlier occurrence should not be surprising. In many algorithmic tasks, and particularly basic ones, the text phrasing includes terms that may "tempt" the problem solver to invoke corresponding computational schemes. This may be particularly relevant with terms like "count the number of elements," "sum the elements," "is there a combination of elements," and the like. Although such phrasings may invoke preliminary associations with supposedly "called for" design patterns, one should be careful, and try to gain sufficient mathematical insight before deciding on the algorithmic design direction. The phenomenon of the students' inappropriate hasty way, versus the desired careful analysis way, is apparent not only in computer science, but also in other domains, such as mathematics (Schoenfeld, 1992).

In the last two game tasks, the students' behaviors reflected the undesired behavior of premature conviction and generalization from simple cases. In our experience, this phenomenon is widely evident with algorithmic problem solvers. Many tend to examine several simple examples and conclude a design scheme. Although this tendency may sometimes yield appropriate algorithmic solutions, it is improper and error prone. One should reach solid conviction of the design scheme. Unfortunately, many students did not seek solid correctness argumentation, or alternatively, evidence that may falsify an unfounded conviction. When confronted with falsifying evidence, they just locally refined, and supposedly "fixed" their initial conviction by some "local patching."

In each of the last three tasks, students' intuitive arguments were unfounded and incorrect. In the powers of 2 task, some were convinced that the only relevant direction was that of searching for a suitable combination of components, among all the relevant combinations. In both of the game tasks, they argued correctness "operationally" by reiterating their strategy rules. They particularly lacked the necessary assertional perspective that yields sound arguments. An operational kind of argumentation may be natural to many, but is insufficient (Dijkstra, 1989; Gries, 1981), as can be clearly seen here.

Moreover, in the last two tasks, they justified the correctness of their solutions with empirical data, namely examples. Studies in mathematics education report on students' empirical proof schemes (Chazan, 1993; Harel & Sowder, 1998). The computer science students in this study were not explicitly asked to prove their solutions, yet they were expected to offer arguments more solid than particular examples or reiteration of rules. Unfortunately, they avoided seeking solid mathematical justification, based on sound mathematical patterns.

Beliefs

In the interviews with students about their solutions, we did not explicitly ask them about their beliefs and tendencies. We extracted our impression from their behavior and comments on their solutions. In our impression, students were unaware, and possibly unconvinced of the essential illuminating role of mathematical patterns in algorithm design. In addition, they seemed to lack a quest for elegance, which was relevant in all the suitable solutions.

It seems that they did not question their ways. This may partly be attributed to their repeated success with designing "working" algorithms, or computer programs. Their algorithms "work" on their test cases, and therefore are "satisfying," without any ordered examination of correctness and efficiency. This behavior reflects unawareness of the significant role of mathematically founded design and analysis. Thus, mathematical awareness should be significantly enhanced. One clear conclusion from this study is the importance of significant enhancement of number patterns, and the elegance they yield in algorithmic problem solving.

CONCLUSION

This chapter presented the phenomenon of overlooking number patterns in solving algorithmic tasks. The overlooked number patterns were not explicit in the task descriptions, yet they could be recognized through careful examination of the task characteristics. Some students performed such examination, but many avoided it. The different sections in this chapter displayed various inappropriate student tendencies that yielded this avoidance. Although these tendencies varied, their outcomes were identical—unsuitable, incorrect or inefficient algorithmic solutions.

The illuminated inappropriate tendencies should be addressed by computer science educators. First, educators should be aware of their presence, characteristics, and outcomes. Then, the teaching of algorithmic problem solving, at all levels, should involve repeated emphasis on mathematical patterns in general, and number patterns in particular. One should embed assertional, mathematical patterns in the design of algorithmic solutions (e.g., Dijkstra, 1976; Gries, 1981; Ginat, 2004). In addition, one should underline the fruitful, desired outcomes of such embedment, in comparison to the undesired outcomes of avoiding it.

Students may, and should also learn from their mistakes. One may let students follow the wrong allies, reach supposedly "working" solutions, and then realize their inappropriateness (Borasi; 1996; Ginat, 2003). Repeated experiences of this kind may enhance the realization of the essential need for mathematical patterns.

These suggestions may not always be simple to implement. Many programming, and even algorithms textbooks do not emphasize mathematical patterns. This might partly be due to an impression that such patterns are not a "must," and therefore may be a burden, because they are not always easy to reach or specify. These textbooks may often create the impression that one can do without them. Computer scientists debate about the role and emphasis of mathematical patterns. Yet, it is evident time and again that the role of mathematical insight is fundamental at all levels (e.g., Bruce, Drysdale, Keleman, & Tucker, 2003; Henderson et al., 2001). The results displayed here strengthen this evidence, and demonstrate the unfortunate outcome of doing without suitable mathematical patterns.

REFERENCES

Astrachan, O., Berry, G., Cox, L., & Mitchener, G. (1998). Design patterns: An essential component of CS curricula. *Proceedings of the 29th SIGCSE Technical Symposium on CS Education*, 153–160.

Borasi, R. (1996). *Reconceiving mathematics instruction: A focus on errors*. Ablex.

Briars, D. J., & Larkin, J. H. (1984). An integrated model of skill in solving elementary word problems. *Cognition and Instruction, 1,* 245–296.

Bruce, K. B., Drysdale, R. L. C., Keleman, C., & Tucker, A. (2003). Why Math? *Communications of the ACM, 46*(9), 41–44.

Chazan, D. (1993). High school geometry students' justification for their views of empirical evidence in mathematical proof. *Educational Studies in Mathematics, 24,* 359–387.

Cormen, T. H., Leiserson, C. E., & Rivest, R. L. (1991). *Introduction to algorithms*. Cambridge, MA: MIT Press.

Dijkstra, E. W. (1976). *A discipline of programming*. Englewood Cliffs, NJ: Prentice-Hall.

Dijkstra, E. W. (1989). A debate on teaching computing science. *Communications of the ACM, 32*(12), 1397–1414.

Fischbein, E. (1987). *Intuition in science and mathematics*. Dordrecht, Netherlands: Reidel.

Fischbein, E. (1999). Intuitions and schemata in mathematical reasoning. *Educational Studies in Mathematics, 38,* 11–50.

Ginat, D. (2001a). Loop invariants, exploration of regularities, and mathematical games. *International Journal of Mathematical Education in Science and Technology, 32*(5), 635–651.

Ginat, D. (2001b). Early algorithm efficiency with design patterns. *Computer Science Education, 11*(2), 89–109.

Ginat, D. (2003). The greedy trap and learning from mistakes. *Proceedings of the 34th SIGCSE Technical Symposium on CS Education,* 11–15.

Ginat, D. (2004). Embedding instructive assertions in program design. *Proceedings of the 8th ITiCSE Conference on Innovation and Technology in CS Education,* 62–66.

Gries, D. (1981). *The science of programming*. Springer-Verlag.

Harel, G., & Sowder, L. (1998). Students' proof schemes: Results from exploratory studies. *Issues in Mathematics Education, 7,* 234–283.

Hegarty, M., Mayer, R., & Monk, C. (1995). Comprehension of arithmetic word problems: A comparison of successful and unsuccessful problem solvers. *Journal of Educational Psychology, 87,* 18–32.

Henderson, P., Baldwin, D., Dasigi, V., Dupras, M., Fritz, J., Ginat, D., et al. (2001). Striving for mathematical thinking. *SIGCSE Bulletin, 33*(4), 114–124.

Linn, M. C., & Clancy, M. J. (1999). Patterns and pedagogy. *Proceedings of the 30th SIGCSE Technical Symposium on CS Education,* 37–42.

Lithner, J. (2000). Mathematical reasoning in task solving. *Educational Studies in Mathematics, 41,* 165–190.

Oprea, J. M. (1988). Computer programming and mathematical thinking. *Journal of Mathematical Behavior,* 7(2), 175–190.

Polya, G. (1957). *How to solve it* (2nd ed.). Princeton, NJ: Princeton University Press.

Schoenfeld, A. (1985). *Mathematical problem solving*. San Diego, CA: Academic Press.

Schoenfeld, A. (1992). Learning to think mathematically: Problem solving, metacognition, and sense making in mathematics. In D. A. Grouws (Ed.), *Handbook of research on mathematics teaching and learning* (pp. 334–370). New York: Macmillan.

Chapter 11

Revisiting Algebra in a Number Theoretical Setting

Jennifer Christian Smith
University of Texas at Austin

Mathematics educators and mathematicians have recently begun to reconsider the university mathematics preparation of teachers, paying careful attention to the development of knowledge of mathematics for teaching: an understanding of the underlying processes and structures of concepts, relationships between different areas of mathematics, and knowledge of students' ways of thinking and mathematical backgrounds. Precisely how this knowledge is to be developed is currently a topic of much consideration and research.

As demonstrated in this book, the study of number theory provides many opportunities for preservice teachers to look for patterns and to make connections between abstract concepts in algebra and arithmetic. It also provides an opportunity for researchers to examine these teachers' understandings of many fundamental mathematical concepts. In my dissertation study, I followed six undergraduate mathematics students through a semester-long number theory course at a large state university. In particular, I paid attention to these students' interpretations and applications of the concept of *congruence of integers*. In this chapter, I discuss aspects of these students' conceptions of congruence and compare them to children's conceptions of similar algebraic concepts.

The course in which the study was conducted was not specifically focused on providing students with opportunities to make connections between undergraduate and secondary mathematics concepts. The results presented in this chapter suggest that having students examine familiar concepts in a new context provides valuable opportunities for instructors to lead the class to reexamine the structure of fundamental concepts in algebra. It may be that carefully guided experiences in novel and unfamiliar contexts, such as those afforded by elementary number theory, can help preservice teachers develop a more sophisticated mathematical perspective, as well as a deeper understanding of the mathematics they will be teaching.

BACKGROUND

The concept of congruence is related to several different areas of mathematics: elementary and abstract algebra, equivalence, divisibility, and division with remainder. In this section, I review and summarize relevant research on the teaching and learning of these topics.

The teaching of algebra is arguably the largest component of the job of a secondary school mathematics teacher. Secondary schools in the United States typically offer three levels of courses in algebra, and most universities in the United States require students applying for admission to have completed at least 2 years of algebra study. In addition, algebra is the foundation for much of the mathematics secondary school students will study. According to the National Council of Teachers of Mathematics (2001), algebra is an "essential component of contemporary mathematics and its applications in many fields" (p. 1). Many researchers have emphasized that, in addition to studying a good deal of mathematics at the undergraduate level, prospective teachers need to develop knowledge of mathematics for teaching, an understanding of the underlying processes and structures of concepts, the relationships between different areas of mathematics, and knowledge of students' ways of thinking and mathematical backgrounds (CBMS, 2001; Fennema & Franke, 1992; Ma, 1999).

It has become clear in recent years that knowledge of mathematics for teaching is not easily developed. For most prospective teachers, there is what Cuoco (2001) called a *vertical disconnect* between the undergraduate mathematics they study and the mathematics they will teach; "this is especially true in algebra, where abstract algebra is seen as a completely different subject from school algebra" (p. 169).

Undergraduates do not necessarily recognize that the topics studied in abstract algebra provide explanations for why certain equations can be solved and others not, nor that they provide rationale for many of the processes of high school algebra (Usiskin, 1988). The Mathematical Education of Teachers (CBMS, 2001) recommends that prospective teachers take courses in abstract algebra and number theory in order to examine the mathematical structures foundational to algebra and number systems, noting that the connections between these may need to be made in other courses. When these connections are not made, teachers rely on their own precollege algebra education when teaching algebra, "an experience that was likely to have been focused on an algorithmic approach to mathematics and unlikely to have contributed to conceptual understanding" (Ball & McDiarmid, 1990, p. 441). This certainly reflects my own experience as a secondary school mathematics teacher.

Zazkis (1999) advocated having preservice teachers reexamine familiar mathematical processes and objects in unconventional number systems as a means to get students to "reconsider their basic mathematical assumptions and analyze their automated responses. [These types of activities] constitute an essential tool for the development of critical thinking in mathematics teacher education" (p. 650). I strongly agree with this perspective and suggest that the study of many topics in abstract algebra and number theory can provide an ideal opportunity to examine teachers' fundamental understandings of algebra. For example, studying the properties of functions and equations in finite rings such as $\mathbb{Z}/n\mathbb{Z}$ could enable students to make explicit connections between and deepen their understandings of the ways in which algebraic structures underlie the processes of secondary school algebra, such as modeling situations with functions and equations, finding roots of polynomials, and using various procedures for solving equations.

Typically, a prospective secondary mathematics teacher attending a university in the United States will complete a sequence of courses analogous to those required for a mathematics major. In some cases, these students take the same undergraduate courses as other students majoring in pure or applied mathematics, and in other cases, some of the upper division courses are designed especially for them and focus on topics related to the high school curriculum. Most teachers encounter abstract algebra and number theory at some point in their undergraduate experience.

Congruence of Integers

Many students in U.S. universities are not exposed to modular arithmetic prior to enrolling in an abstract algebra or number theory course. Although not a part of the traditional elementary or secondary mathematics curriculum, the topic is often introduced in computer programming courses as "mod (a,b)," a function of two variables. Although this certainly has an impact on some students' understanding of this concept when they later encounter it, I was primarily interested in studying students' conceptions of congruence as an equivalence relation in the context of a mathematics course.

In our daily lives, we experience this idea in the form of "clock arithmetic." For example, if it is 10:00 now, and we would like to know what time it will be in 15 hours, we can notice that adding 12 hours gets us back to 10:00

again, and adding 3 more hours brings us to 1:00. We see that adding 3 hours is the same as adding 15 hours, in terms of the relative positions of the hands on the clock. In the language of congruence, we have that $3 \equiv 15$ (mod 12). There are many variations on this theme, such as deciding on which day of the week an event n days away will fall.

Arabic mathematicians were familiar with arithmetic of remainders a millennium ago. According to O'Connor and Robertson (2002), a Persian mathematician named ibn al-Haytham (pp. 965–1040) solved problems involving remainders using what is now called Wilson's theorem.[1] One thousand years earlier, the Chinese were able to solve systems of linear congruences, although these were phrased in terms of remainders left on division by various integers.

Although mathematicians had been doing arithmetic with remainders for millennia, it was Gauss who introduced the concept of *congruence* in the opening sections of his 1801 work, *Disquisitiones Arithmeticae*. According to Ore (1948), Gauss defined this relation as follows: "Two integers a and b shall be said to be *congruent for the modulus m* when their difference $a - b$ is divisible by the integer m" (p. 211). Gauss also introduced the modern notation $a \equiv b$ (mod m), and gave the following examples of congruences in *Disquisitiones:*

$$16 \equiv -9 \pmod{5}$$
$$-7 \equiv 15 \pmod{11}$$
$$-7 \not\equiv 15 \pmod{3}$$

Note that in these examples, there is no canonical form for the integers in each congruence; that is, Gauss did not insist that the integer on the right side of the congruence be the standard congruence class representative.

Modular arithmetic and congruence are important concepts in modern mathematics. They generally appear in the first chapter of an undergraduate number theory text and are simply assumed to be understood in many mathematics texts. Congruence also plays an important role in the undergraduate curriculum as an example of a nontrivial equivalence relation. The system of *residues* modulo an integer n is an important example of a quotient ring, and is (assumedly) fairly accessible to students in abstract algebra. The relation of congruence modulo n partitions the integers into classes; we can then operate upon these as objects in a finite number system. This is generally presumed to be a good introduction to the following more general mathematical process: Take a set of objects, along with its mathematical structure; choose a certain subset of these objects; use an equivalence relation to somehow "mod out" by this sub-

[1]Wilson's theorem states that for p prime, $p \mid ((p-1)! + 1)$.

set; and then consider the resulting quotient, which tends to be another set with a related type of structure. It often is the case that this new quotient object has certain properties that make it a powerful setting in which to solve certain mathematical problems from the original set. This idea of *translating* problems to other domains in order to facilitate their solution is a major approach to solving problems in nearly all fields of mathematics.

Abstract Algebra

The course in which the research reported in this chapter was conducted was envisioned as a course covering abstract algebra and number theory relevant to teachers. Although the degree of formalism one would expect in a standard abstract algebra course for mathematics majors was not always present in the course studied, it is important to note that many of the participants were making sense of abstract mathematical concepts for the first time. They struggled with many of the same issues as those reported in the research on the learning of abstract algebra.

Dubinsky, Dautermann, Leron, and Zazkis (1994) wrote that "abstract algebra is the first course for students in which they must go beyond learning 'imitative behavior patterns' for mimicing [sic] the solution of a large number of variations on a small number of themes (problems)" (p. 268). For many students, this course marks their first experience with abstract mathematical concepts. The concepts are not only introduced abstractly (in terms of definitions and properties), but are then studied further by examining "what facts can be determined just from [the properties] alone" (p. 268). This is a new experience for most students, and they struggle a great deal with this new way of learning and thinking about mathematics.

Dubinsky et al. (1994) noted that "constructing an understanding of even the very beginning of abstract algebra is a major event in the cognitive development of a mathematics student" (p. 295). Although no specific studies have been done, many researchers have remarked that students who struggle in an abstract algebra course may become reluctant to continue to study abstract mathematics. Dubinsky et al. (1994), Hazzan (1999), and Clark et al. (1997) argued that an abstract algebra course is of critical importance to mathematics students, for this is the course in which they develop many of the abilities they will need for future study in mathematics. Because a significant proportion of mathematics majors will become high school teachers, this course plays a critical role in developing teachers' knowledge of and attitudes toward mathematical abstraction (Clark et al., 1997; Dubinsky et al., 1994).

Elementary Number Theory

In the last few years, a growing body of research on the teaching and learning of elementary number theory has emerged. Much of this work seems to have been inspired by Zazkis and Campbell's (1996) study of preservice elementary teachers' difficulties with the concept of divisibility. For example, students often confuse multiples and divisors of a number, and generally have difficulty reasoning about divisibility when integers are presented in factored form. Zazkis and Campbell also noted that the students in their study demonstrated a *procedural disposition* toward determining divisibility, and toward finding greatest common divisors and least common multiples.

Brown, Thomas, and Tolias (2002) reported that students tend to view a factored form of a number as an instruction to calculate the product, rather than as an alternate representation of the numbers; an interesting example of this phenomenon was given by Zazkis and Campbell (1996). Given $M = 3^3 \cdot 5^2 \cdot 7$, students were asked to determine if M was divisible by 7, 5, 2, 9, 81, 63, and 15. One student noted that 7 and 5 were divisors of M, because these appeared in its prime factorization. Similarly, he decided that 2 was not a factor since it did not appear in the factorization. Bob concluded that 9 was a divisor, reasoning that because M was a multiple of 3, any multiple of 3 less than M would also divide M. When asked about 81, 63 and 15, he decided the only way to determine divisibility was to calculate M by expanding the product, and then dividing the result by 81, 63, and 15. Clearly Bob did not see the power of the prime factorization representation of the number M.

Several researchers (Ferrari, 2002; Rowland, 2002; Selden & Selden, 2002) have remarked that an elementary number theory course is an excellent setting in which to study students' reasoning and proof strategies. Because the topics studied in a number theory course typically assume only knowledge of arithmetic, it is thought to be a course in which students can learn to reason about mathematical ideas. Rather than struggling with the concepts and content, they can focus on proof, justification, and problem-solving strategies. Ferrari (2002) studied undergraduates' use of representations in a number theory course and found that students have difficulty distinguishing between a number, its representations, and related properties. For example, students were asked, if $M = 3^4 \cdot 5^3 \cdot 7^6 \cdot 19^8$, is $M + 5$ divisible by 10? This is a question that cannot be answered simply by looking at the prime factorization of M, and the number was chosen to be too large to facilitate the use of a calculator. In other words, students must reason about divisibility in order to answer this question.

In a 1998 study, Zazkis examined prospective elementary teachers' conceptions of even and odd numbers. She found students did not view divisibility by 2 and evenness as being equivalent properties of a number. Being even was seen as a function of the last digit of a number, whereas determining divisibility by 2 seemed to necessitate that a division operation take place. For example, one student struggled to construct an example of a five-digit number that left a remainder of 1 upon division by 2, although she had been able to give one- and two-digit numbers with this property. Another example particularly germane to the topic of this chapter concerns a student asked to determine if the number $K = 6 \cdot 147 + 1$ was divisible by 6. The student initially responded that 6 did indeed divide K, claiming that 6 appeared in the prime factorization. When pressed to test this conjecture with a calculator, the student realized she had been incorrect. She then explained that 6 did not divide K because K is odd and 6 is even. It is interesting to note the student did not recognize K would leave a remainder of 1 upon division by 6. This, along with other similar results, suggests that students do not easily see the connection between division with remainder and divisibility.

Campbell (2002) argued that students' difficulties with concepts like divisibility and division with remainder were the result of "inappropriate extensions and adaptations of more common and familiar procedures and concepts" (p. 32). He reported an instance of a preservice teacher expressing division with remainder inappropriately as an equation [A ÷ B = Q rem R] and then algebraically transforming this equation into one similar to the division theorem [A = BQ + R, where $0 \leq R < B$] by simply replacing the "rem" with a "+."

Equivalence

Because congruence is an equivalence relation, we might expect to see parallels between understanding of congruence and understanding of other equivalence relations. There is a large body of research on K–12 students' interpretations of equality dating back to the 1970s. The equality symbol "=" is inherently vague and can have different meanings depending on context (Freudenthal, 1983); this is due largely to the somewhat chaotic development of mathematical symbolism throughout history (see Saenz-Ludlow & Walgamuth (1998) for an interesting account). For example, = can be a *defining symbol,* as in $f(x) = x^2 + 2x - 3$, an *operational* symbol indicating the result of a computation, such as $f(-3) = 0$, or a *relational* symbol, indicating two quantities have the same value, as in $f(-3) = f(1)$. Ideally, we would like students to be able to interpret the equality symbol appropriately in each context. Behr, Erlwanger, and Nichols (1980) reported that children typically view the equality symbol exclusively as an operational symbol; a command to do something. When presented with statements like $8 = 3 + 5$, first grade students responded that these were wrong, and would switch the positions of the "=" and the "+" to get $8 + 3 = 5$, and then note the 5 should be an 11. Second graders presented with statements like 3 + 2 = 2 + 3 also tended to claim these were false, because 3 + 2 is 5, not 2.

Difficulties interpreting the equality symbol are not confined to elementary school children. Kieran (1981) and Wagner (1977) showed that secondary school students typically regard the equality symbol operationally—as a "unidirectional symbol preceding a numerical answer" (Booth, 1988, p. 24), instead of relationally—indicating two quantities are the same. This evidence suggests children do not necessarily "grow out of" this interpretation of equality as they progress through school. Kieran (1999) reported a successful teaching experiment in which children's conceptions of equations were developed through a carefully designed and implemented sequence of instructional activities that built upon the children's existing operational understandings of equality. At the beginning of the experiment, all of the participants exhibited an operational interpretation of equality, and the activities eventually led the students to view algebraic equations relationally. It is interesting to note that these were seventh- and eighth-grade students, and not elementary school children.

Research on children's interpretations of algebraic equations and the process of solving these equations reveals there are many conceptual difficulties. According to Booth (1988), "in algebra, the focus is on the derivation of procedures and relationships and the expression of these in generalized, simplified form" (p. 21). Students have difficulty accepting algebraic expressions as answers, preferring to pick values for the variables in order to give a numerical answer. Kieran (1988) reported that when solving equations, beginning algebra students tended to rely on a memorized procedure that appeared to disregard the role of the equality symbol in the equation. Wagner and Parker (1999) described the difficulties students with an operational view of equality often face when solving equations in algebra; most solution methods assume a relational view of the equality symbol, and so students must work with the entire relation as they transform it into equivalent relations. They noted that "few students fully appreciate the fact that solving an equation is finding the value(s) of the variable for which the left- and right-hand sides are equal" (p. 333). According to Bernard and Cohen (1988), understanding how to solve equations by the equivalent-equations[2] procedure is a conceptually sophisticated task that requires a good deal of cognitive preparation. They claim methods typically used in prealgebra, such as guess-and-

[2]The *equivalent equations* method of solving equations is the one in which one transforms an equation into an equivalent, simpler representation by performing some allowable transformation on the equation, such as adding the same number to both sides of the equation, or applying the sine function to both sides of the equation. After a series of such transformations, the solution to the equation becomes clear.

check, the "cover-up" method (viewing equations as arithmetic identities with one value covered up), and the "undoing" method (viewing equations as a sequence of reversible steps that have been applied to a number), although important activities, are not adequate preparation for learning to solve equations using the equivalent-equations method. Note that a student with an operational view of equality can successfully learn to solve simple equations by such methods, because a relational view of equality is not necessary in order to guess the value of the missing number and then perform arithmetic operations to check if the result is correct.

Herscovics and Kieran (1999) also reported that students have a great deal of difficulty solving equations using the equivalent-equations procedure. Kieran (1999) stated that although research shows many students become adept at solving equations in an automatic, procedural fashion, "these studies demonstrate that the same students are generally not aware of the structure underlying the manipulations they perform" (p. 351).

METHODS

The results presented here are from the author's dissertation study (Smith, 2002). An exploratory case study design was used to investigate undergraduates' conceptions of congruence of integers in an introductory abstract algebra and number theory course at a large state university in the southwestern United States. The researcher's role was that of a participant–observer and teaching assistant in the course.

Participants

The six participants were chosen based on high scores on a first exam, in conjunction with Dr. Thomas, the instructor of the course. We chose students whom we perceived to be hard-working, conscientious mathematics students with the potential to excel in the course. In addition, the students were chosen to represent as diverse a group from the class as possible, representing both men and women, and "traditional" and "nontraditional" students. Each had earned a grade of A or B on the first exam, and all eventually earned either an A or a B in the course. The six participants' pseudonyms (Andrew, Barbara, Chris, Dan, Eva, and Fran) indicate their relative exam averages in the course; that is, Andrew had the highest exam average of the six, whereas Fran had the lowest. Except for Barbara, who went directly into a graduate program in mathematics education, the participants all became high school mathematics teachers after finishing their degrees.

Nature of the Course

The course, titled "Introduction to Algebra and Number Theory," was a junior-level (third-year) course and required for prospective secondary mathematics teachers, although it was often taken as an elective by pure mathematics and computer science students as well. The course was originally conceived as a "watered-down" abstract algebra course for teachers. In the late 1980s, Dr. Thomas began teaching the course and decided to emphasize number theory and problem solving over an axiomatic treatment of groups, rings, and fields.

Although the class was informal and interactive, Dr. Thomas did not tend to use group work in class or have the students discover ideas on their own. He did not lecture either; each class could best be described as a conversation between Dr. Thomas and the students. He did most of the talking, but he asked a lot of questions as well. He valued understanding, exploration, and intuition alongside mathematical rigor, and modeled all of these for the students. When the class was working on a problem, about half of the students were typically listening intently and offering suggestions, while the other half, bent over their notebooks and calculators, worked on the problem on their own.

Dr. Thomas attempted to build the concept of congruence on the students' prior knowledge of division. He expected them to realize that two integers are congruent modulo n if they leave the same remainder upon division by n, and that this idea leads to the standard definition of congruence: $a \equiv b \pmod{n}$ if and only if n divides $a - b$. He wanted the students to make sense of congruence and come to this definition because it made sense, not simply because it was defined this way.

I should note at this point that the focus of this study was on the six participants and how they made sense of congruence in this particular context. Dr. Thomas and his perspective on teaching congruence were not being investigated. As the instructor of the course, Dr. Thomas undoubtedly had an impact on the students' conceptions, but the focus here is on the participants' thinking about the concept. I encourage the reader to keep this point in mind while reading this chapter.[3]

Data Collection

The six participants completed questionnaires and participated in three semi-structured interviews over the course of the semester. The first inter-

[3]I have found in presenting these results that the questions asked by some mathematics faculty often focus on *how* the topics were presented, *what* the instructor did, and *why* various ideas were or were not emphasized, as if the students' misconceptions and difficulties could be easily "fixed" by simple saying the right words. This perspective not only trivializes this type of research, but also the complexity of both the profession of teaching and of the process of learning mathematics.

view took place one week prior to the exam covering congruence of integers, a second interview took place approximately 2 weeks later, and a third approximately 3 weeks after the second. In addition, class sessions during the unit covering congruence and applications of modular arithmetic were videotaped and transcribed. Data were analyzed using a grounded theory approach (see, for example, Strauss & Corbin, 1998).

Interview Tasks

The results presented in this chapter come from questions asked in the interviews concerning written questionnaires the participants had completed in advance. The following items appeared on one of the questionnaires:

1. How would you explain what the statement $a \equiv b \pmod{n}$ means?
2. Is there another way you can think of to explain it? (If so, please do.)
3. Can you think of a way to draw a picture to show what this expression means?
4. What values could x have in each of the following?
   (a) $x \equiv 3 \pmod{7}$
   (b) $17 \equiv x \pmod{4}$
   (c) $x \equiv 23 \pmod{11}$
   (d) $2 \equiv x \pmod{9}$

During the interview, each participant was asked to explain and expand upon his or her written answer. In addition, the students were asked to discuss their responses to the following items on the unit exam:

- Solve for x: (a) $5x \equiv 1 \pmod{11}$
  (b) $6x \equiv 3 \pmod{15}$
- $19^{19} \equiv \_\_ \pmod{7}$

RESULTS

Two major findings of the study were that students tended to view the relation of congruence operationally rather than relationally, similarly to the way that children tend to view the relation of equality, and that the participants demonstrated a tendency to solve linear congruences using a complicated memorized procedure, instead of using the properties of the congruence relation and the algebraic structure of the finite rings of integers in which the congruences existed. Both of these are reminiscent of children's difficulties when learning to solve linear equations in algebra.

Operational and Relational Conceptions of Congruence

Earlier in the chapter, I briefly discussed the research on children's interpretations of the equality symbol. This type of research was done not by pointing to an equality symbol and asking children to explain what it meant, but by carefully examining children's interpretations of equations and arithmetic identities. It became clear early in the analysis that the participants in this study were not viewing congruences as equivalence relations, but as statements about a computation or a transformation that had occurred. I began to believe that the students in this study were interpreting congruences in a similar way as children frequently interpret equality.

I first more precisely define[4] what is meant by operational and relational interpretations (or views) of congruence. Having an operational view of the congruence $a \equiv b \pmod{n}$ means interpreting the statement in terms of a completed operation, with the number on the right as the "answer." Having a relational view of congruence means interpreting the statement as a relationship between two equivalent quantities. Table 11.1 lists the indications of each view.

Interpretation of Congruence. After the students had learned to use the fact that integers can be classified in various ways to solve problems (i.e., all integers are of one of the forms $3k$, $3k + 1$, or $3k + 2$), the notation "$a \bmod n$" was introduced as a way to express the idea that a number is of the form $a + kn$, and the notation "$a \equiv b \bmod n$" to denote that two integers are of the same form. Dr. Thomas asked the students to determine when two integers could be considered congruent modulo n, and the following three definitions emerged from the class discussion:

1. a and b are congruent mod n if and only if they leave the same remainder when divided by n.
2. a and b are a multiple of n apart, and so we can write $a = b + nk$
3. $n \mid a - b$

The third definition is the standard definition for congruence found in most textbooks. Dr. Thomas explained that each of these definitions was equivalent to the other, and after this episode, no further attention was given in class to the interpretation of congruences.

It is therefore interesting to note that when the participants were asked in an interview to explain the meaning of the statement $a \equiv b \bmod n$ in at

[4]These definitions were developed from the data as I attempted to separate the students into these two categories. As I questioned and revised my thinking, returning to the data frequently, the definitions developed.

TABLE 11.1
Operational and Relational Views of Congruence

Operational	*Relational*
Interpreting congruences frequently, if not exclusively, as statements like a / n leaves a remainder of b	Interpreting congruences in a variety of ways, but the division remainder interpretation implies that a and b leave the same remainder upon division by n
View reducing right side of a congruence as a requirement for the congruence to be true	View reducing the right side of the congruence as a matter of convention, or as a problem-solving device, depending on context
Viewing congruence as a transformation from $\mathbb{Z}$ to a "mod n world"[a]	Viewing congruence, formally or informally, as an equivalence relation defined on $\mathbb{Z}$, so that the quantities on both sides of the congruences are seen as elements of the same set
Viewing the mod n term as attached to one of the integers in the congruence	Viewing the mod n term as modifying the entire congruence

[a]The students in the course referred to the sets of residue classes modulo n as the "mod n world." Their use of this term is discussed later in the chapter.

least two ways, all of the participants except Chris gave the following interpretation first; "a divided by n has a remainder of b." When asked in the first interview if she could think of another way to explain what a congruence statement meant, Fran attempted to rewrite the congruence statement as an equation as she had seen Dr. Thomas do in class, but only managed to write $n = a - b$. At that point, she seemed to be confusing the standard definition with the rewriting of a congruence as a linear equation of the form $a = b + nk$.

Eva's interpretation of the congruence statement ("if I take n and I multiply it with a whole number and then I add b to it, I should get a") is an example of her tendency to think about congruent integers in terms of a process of transforming one integer into another by adding or subtracting multiples of n. Note that interpreting a congruence in this way does not imply understanding that congruence indicates a relationship between two quantities. Rather, the symbol $\equiv$ indicates that the operation of division has occurred, with the remainder noted. This is similar to children's operational interpretations of equality.

Dan had developed an informal notion of congruence classes, which he called families. He demonstrated using the modulus $n = 6$, writing down zero through 5 and then noting that all integers fall into one of these families. He appeared to be thinking of 0, 1, 2, 3, 4, and 5 in this sense as congruence class representatives; each was the representative of the number family in question. "I look at a is congruent to b mod n as meaning a is just some number, and ... you can say n divides a with a remainder of b or you can say that a belongs to the b mod n family, or b is within [the range] zero to n." Note that this notion of classifying numbers also indicates an operational interpretation of congruence. Dan is not saying that a and b are in the same family; he is saying that a is in the b family, where b is considered the standard representative.

Having a relational interpretation of congruence would entail understanding that the symbol ≡ indicates that two quantities are considered the same, that their difference is divisible by the modulus, that upon division by the modulus, both numbers have the same remainder upon division by n, and so on. For example, when asked about the meaning of congruence in the first interview, Chris wrote down the equation $a = b + nk$, and then gave the following explanation:

> I guess I think of a and b as the *same number* under this mod. They're the same number. They're in the same equivalence class, and that means there's an n separation between them, and that could be a zero, it could be whatever n is. So $b - a$ could be n or $a - b$ could be n or it could be $2n$, or $3n$, or $4n$, or $5n$.

In the quote, Chris gave three more meanings of congruence: Congruent integers are regarded as the "same," congruent integers fall into the same equivalence class, and the difference between any two congruent integers is a multiple of the modulus. Each of these emphasizes a relational interpretation of congruence.

Reducing the Right Side. At the time of the first interview, several of the participants expressed a preference for reducing the integer on the right side of a congruence to the smallest positive integer smaller than the modulus. Eva reduced what she called "ugly" numbers (e.g., negatives) saying, "I don't like negative numbers [on the right]. And I think Professor Thomas is the same way because he always changes the negative numbers to positive." Dan noted that the reduced numbers were more comfortable to work with. "If we see a number bigger than n, or equal to n, we want to bring it back down to a more comfortable level .... Negative 2 is in the family of 4 [modulo 6] and 4 is within my comfortable family range, being 0, 1, all the way up to 5."

Chris and Barbara both recognized that it was often useful *not* to reduce the integers on the right. Chris said, "I don't tend to think of the negative numbers, although [Dr. Thomas] did an example in class the other day that was so much easier doing the negative numbers, so I'm starting to change that a little bit more." Barbara stated that she had no preference for numbers to be of a particular form on either side of the congruence. "You could write like ... a is congruent to 4 mod 4. Like you could have a situation where it's not totally reduced .... There are definitely times when you would want the smallest integer number. Although I don't think I can think of one right away."

Both Fran and Andrew seemed to be thinking of congruence largely in terms of division with remainder, and they both viewed the integer on the right side of the congruence as *the* remainder. Neither of them made any comments during the first interview that indicated a belief that the right side of the congruence should always be reduced, or that this was not necessary. By the second interview, Andrew had thought about this issue, whereas Fran had not.

The participants seemed more inclined to reduce the right side of a congruence by the second interview, although they were uncertain whether or not this was important. Andrew and Eva seemed to think that reducing the right side was a matter of convention, but a necessary one, similar to the arithmetic practice of reducing fractions to lowest terms. Both had given unreduced answers for a problem on the exam [$19^{19} \equiv$ __ (mod 7)],[5] and subsequent feedback from the instructor had convinced them that these should have been reduced. In response to losing points for his answer of 19 for this problem, Andrew said, "I don't know. I guess in some sense it's sort of a matter of convention. It's a very minor thing and I'm not really too bogged down by it. I mean, as far as I can think, I mean, it's still right and, you know, I mean the congruence is still true just like two fourths is still one half." Eva also used the fraction analogy, saying, "It is the same thing like one half or you have like two fourths. These represent the same thing, but I like one half."

Dan believed one should always reduce the right side of a congruence. Asked if 19 would have been a correct answer to the above problem, he said, "You know you can reduce it further since 19 is not within the range of 0 to 6. I knew I could reduce it, so I would have never ... I wouldn't have left it like that." It seemed that he had never questioned whether or not one *had* to reduce the number on the right; he simply always did.

Barbara and Chris both seemed less certain about this issue than they had been in the first interview. Barbara was not as adamant as she had been earlier that one could write any integer on either side of a congruence. "See that's what's confusing to me is like I've seen problems where this [right side] is greater than n, but like that just makes me want to reduce it and sub-

[5]See the Appendix for a discussion of the solution of this congruence.

tract a factor of n from it, just so I can see clearly what the remainder is in the mod n world." Chris suggested that perhaps this tendency to reduce the right side of the congruence was the result of seeing most congruences written that way. "I think it's because it's on that side of the congruence. We only see one number on that side and we think of this as being … we tend to write this [right side] as our reduced number and this one [left side] can be anything. And then we don't write mod on both sides of the congruences."

Transformation from $\mathbb{Z}$ to the "Mod World." Many of the participants in this study seemed to have come to terms with the varying interpretations of congruence by combining them into a notion of congruence as a transformation from $\mathbb{Z}$ into what they called the "mod n world." Many of them viewed the numbers or quantities on the left side of a congruence as being ordinary integers, whereas the numbers on the right side of a congruence were regarded as living in the mod n world, and were considered remainders, or families, or congruence classes, or just numbers between 0 and $n - 1$. However, the numbers or quantities on the left were never regarded as one of these.

The term "mod n world" was used in class by Dr. Thomas on the first day that congruence was introduced. By this he meant $\mathbb{Z}/n\mathbb{Z}$, the set of residues of the integers modulo n. The students used this term as well, some more than others, although it was not always clear exactly what the students meant by "mod world." In using this term, Dr. Thomas was trying to avoid introducing formal structures such as rings or fields. He preferred to give the students an intuitive understanding of what another number system was like by using the metaphor of being in a different world. The fact that the students had difficulty understanding congruence probably has little to do with their use of the term. However, the way that this nonstandard term was integrated into their understanding of congruence was very interesting.

Fran and Andrew used the term most frequently. Fran often described working with congruences as working *in* the mod world, whereas working with equations was described as working *outside* the mod world. She preferred to avoid the mod world as much as possible, saying, "Mod world is very confusing to me for some reason and I don't know why." Andrew struggled with the fact that numbers in the mod n world were somehow supposed to be between 0 and $n - 1$, although that rule did not always have to be followed. "It does not make sense to talk about 5 in the world of mod 4 'cause you only consider 0, 1, 2, 3." But when solving congruences, he made substitutions using numbers that did not exist in the mod world. "It's weird to talk about really large numbers that aren't in the world of mod 8, but you have to use that to solve some of these congruences." He seemed to consider the mod world as a separate number system.

The other participants primarily used the term when they were reducing numbers modulo n. For example, Eva asked herself, "what is negative 6 in a mod 8 world?" Dan tended to think of congruence in terms of classifying numbers, saying, "19 to the 19th is in mod 7 world congruent to 1, and every number in the world falls into one of those categories in the world of mod 8." Barbara said, "I would say that in a mod five world, it's four." Chris was the only participant who never used the term "mod world" at all.

In the first interview, only Fran, Eva, and Dan seemed to be thinking of congruence as a transformation. Fran and Eva viewed the numbers on the right side of a congruence as remainders. Fran said, "This [left side] ... it can vary because it's any number, and then this number [right side], I look at it as it should be a number between zero and the mod number because it's a remainder." Eva, when comparing the congruences $17 \equiv x \pmod 4$ and $x \equiv 17 \pmod 4$, said "I don't see them as the same. Because it does matter if you say 17 is the remainder or 17 is the number that you actually started with." In both cases, this transformational view seemed to be based on their fairly limited understanding of congruence at the time; both Fran and Eva tended to interpret congruences primarily in terms of division and remainder.

Dan had a fairly well-developed conception of congruence as a transformation at the time of the first interview. This was probably due to the fact that he viewed congruence in general as a classification of integers, and tended to interpret specific congruences likewise.

> Really, the only numbers that exist in mod 6 are 0, 1, 2, 3, 4, and 5. You know, everything else is just another way to write one of the numbers that does exist. So, what I did here is I converted these all into the numbers that exist in world mod 6 or whatever world we're in.

When comparing the same two congruences as Eva did earlier, he said, "A lot of things are congruent to one mod 4, all right? But not as many things are congruent to ... Seventeen, when divided by 4, there's only one remainder, there's only one answer."

Andrew viewed relative position of integers in a congruence as important at the time of the first interview, and explained the difference between solving the two previous congruences: "you're talking about remainders here [in $17 \equiv x \pmod 4$] and you're talking about finding numbers here [in $x \equiv 17 \pmod 4$]." This was the only comment he made during that interview that indicated a view of congruence as a transformation. He also interpreted congruences in ways that were not indicative of a transformational conception, and so I do not attribute this conception to him. The other participants made no comments during the first interview indicating they were thinking of congruence in this way, but by the time of the second interview, all of the participants showed signs of interpreting congruences as transformations.

At the time of the second interview, Fran was not as reliant on using the word *remainder* as she had been before, but much of the language she used in interpreting congruences indicated that she had a transformational conception:

> Whenever you're living in this world, you're only going to have 10 possibilities to put there [mod 11] basically to be the smallest form.... I mean, there's going to be a bunch of different ways to come up with one of those numbers [left side], but for this one [right side], you just have one.

Eva seemed to have backed away from thinking of congruence as a transformation, but she made occasional comments that indicated this view did influence her thinking. For example, she solved the congruence $5x \equiv 1 \pmod{11}$ using the reduction of moduli procedure on the exam,[6] and she rewrote the congruence $11a \equiv -1 \pmod{5}$ as $a \equiv 4 \pmod{5}$. She understood why she had exchanged the -1 for the 4, but could not explain why she had exchanged the $11a$ for the a. When asked if she could solve this congruence by guessing, she expressed concern about finding all of the possible integers that could be substituted for the x on the left side. However, when solving a similar congruence where the missing number was positioned on the right, she said she only needed to write one answer, because "if you put a 1 here, that's already obvious that it gives you the rest of them very easily. But in [$11a \equiv -1 \pmod{5}$], if you just put a 9, that would not be as obvious."

Dan's interpretation of congruence as a transformation had only strengthened by the second interview. He interpreted most congruences in a variety of contexts in this way. For example, when solving $5x \equiv 1 \pmod{11}$, he said:

> Now on these ones, we've got the x on the outside. It [1] is a family. You're saying basically, 5 times what number will bring you into this family. So a lot of numbers are in that family and that's why it's an infinite sequence. But when you're given a number and you want to classify it into a family, the number only belongs to one family.

Chris and Barbara seemed to be wavering with respect to this interpretation. It was difficult to determine whether their comments were indicating that they believed relative position was important, or if they actually were viewing the congruence in terms of a transformation. Chris seemed to understand the concept of congruence classes, but by the second inter-

[6] See the Appendix for an explanation of the reduction of moduli procedure.

view, his understanding of congruence had shifted toward being transformational as well. For example, while explaining how he approached solving $5x \equiv 1 \pmod{11}$, he said, "I think of this [left] side of the congruence as being any possible number and this [right side] is the class of number it is, it's a 1 mod 11." Barbara had given no indication that she viewed congruence as a transformation in the first interview, but in the second, she occasionally made comments that leaned in that direction, such as, "If the number is next to the mod, it's always going to be a remainder whereas this [other side] can be any number." However, each made only a few comments that indicated a transformational view, and these were all in the context of solving a particular linear congruence; it may be that this interpretation was only influential when faced with congruences such as the earlier one.

By the second interview, Andrew seemed to interpret congruence as a transformation much of the time. This is surprising because he seemed to have decided that position was not important at that time. When explaining why he thought that there was only one answer to the problem $19^{19} \equiv$ __ (mod 7), he said, "$1 + 8k$ could not be the answer because it wouldn't make sense in mod 8. I mean because if you specify 1 plus $8k$, well, if you let k run free, and it's not 0, you're already getting numbers that are not going to be in mod 8."

One phenomenon that I found extremely interesting is related to this last comment. When solving the congruence $5x \equiv 1 \pmod{11}$, all of the participants believed that the correct answer was $x = 9 + 11k$. None of the six participants in the study gave a single number as the answer. In contrast, all gave a single number as the answer to $19^{19} \equiv$ __ (mod 7). They explained their reasoning for this in various ways, but it is striking that they all interpreted the meaning of the variable in the congruence in the same way, and that this was significant in determining the form of the answer. They all seemed to believe that there were infinitely many values for an x on the left, but it was only necessary to give one value for a variable on the right.

The "Mod n" Term. Several of the participants viewed the "mod n" term in a congruence as attached to an integer, rather than as modifying the entire congruence statement. At the time of the first interview, Dan was the only participant who clearly held this view; he would often write congruences with the mod term on the left, such as 3 mod 7 $\equiv x$.

By the second interview, Dan, Eva, and Barbara all seemed to view the mod n term as attached to an integer in the congruence. In many instances, they rewrote congruences as equations by replacing "mod n" by "$+ nk$" in a way similar to Campbell's (2002) division with remainder example cited earlier in the chapter. Eva would occasionally write congruences with the mod term on the left. When asked about this, she said, "That would make more sense to do it that way, because it's obvious that you have a bunch of steps on one side of equals to a number. Like you always write from left to right, you see what I mean?" Barbara did not write congruences this way, but indicated that the number next to the mod term had a certain form. If this guy is next to the mod, then I know that he's going to be a remainder. It's going to be between 0 and n minus 1. But this guy, he can do whatever he wants, I mean being on the left side away from the mod n. She did note however, that though she had never seen congruences written with the mod n term on the left, she saw no reason why it could not be done. "This seems totally commutative to me."

Dan's view was firmly established by the second interview. I asked why he wrote the mod term on both sides of a congruence, and he replied, "I don't feel like it has to be on the right or left. And you can actually see that at this point I switched it so it would be the normal way, but I don't have a problem with writing it on either side." He generally interpreted the number next to the mod term as the "family" of the number, whereas the number on the other side of the congruence was an integer. He definitely viewed congruence as a transformation, but he did not seem to think of the transformation as unidirectional; it could be written with the "family" on the left as well.

It was unclear whether or not Andrew, Fran, and Chris held this view. None made any comments that indicated that they interpreted the mod term as attached to one of the integers or as attached to the entire congruence. Overall, there were no comments made by any of the participants that indicated the latter, which is consistent with their tendency toward an operational view of congruence.

Applying the Concept of Congruence

After being introduced to the concept of congruence in class and learning to solve linear congruences by a guess and check method, the students learned how to use congruence to solve some classic problems in number theory, such as finding solutions for linear Diophantine equations and solving systems of congruences. The students had previously learned to solve linear Diophantine equations (equations of the form $ax + by = c$, where $a, b, c \in \mathbb{Z}$) graphically, by guessing, and by using the Euclidean algorithm. Reduction of moduli (ROM) was then introduced as a means to find all solutions to these equations.[7]

The ROM procedure appears straightforward and powerful. With larger numbers, it can be computationally intensive, but one would expect students with a solid understanding of congruence to understand and be able

[7]See the Appendix for a sample solution.

to use this procedure to solve such equations. However, the participants viewed ROM as a very complex, highly computational procedure. For example, Fran said, "To me it's an algorithm. I'm sure that there's a theory that's behind it that makes a lot more sense but I don't understand the theory." Although the students struggled to understand ROM and were often frustrated with the fact that small computational errors would result in disaster, they quickly appropriated this method and tended to use it whenever they could, including solving linear congruences, much to my surprise.

Solving Linear Diophantine Equations. The students applied ROM to solve the following problem on the exam: Find the smallest x and y such that $75x + 27y = 12$. Note that ROM is a reasonable solution technique for this problem, and this was the method Dr. Thomas expected the students to use for this item.

Fran's explanation of ROM was very procedural, consisting of a series of manipulations and descriptions of precise steps. For example, she rewrote the original equation as a congruence: She started with $75x + 27y = 12$, and then wrote $75x = 12 - 27y$. However, at this point, she was not sure if she could rewrite this as $75x \equiv 12 \bmod 27$, because this was a negative $27y$. At the end of the process, she had found that $x = 7 + 9m$ and $y = -19 + 25m$. She substituted 0 for m to get $x = 7$ and $y = -19$, and was immediately frustrated; these answers had not been awarded full credit on the exam, because this is not the smallest pair of integers that solve the equation. The method of getting the smallest answer appeared quite elusive to many of the participants, and in this case Fran clearly did not view the above expressions as a set of correct solutions. This was a recurring theme among the participants.

Dan was also frustrated by the fact that he did not get the correct answer using ROM on the exam problem. "It's supposed to work that the answers you get are right, and then for some reason there's always some sort of additional trick you're supposed to do to get the smaller answer. And I don't understand where the smaller answer comes from." He explained the process of ROM clearly, although he still seemed to view it very procedurally. I asked if he saw a relationship between the two solutions that he had found, and after a moment of thought, he said, "The difference is 9 ... so it's mod 9. So it actually doesn't ... there must be a way to bring it out, the solution, that it's 7 mod 9." It seemed that although he was aware that there was more than one solution to such an equation, he had not realized that ROM could be used to produce an expression for all of the solutions. In addition, despite the fact that he used congruences to solve this problem, he was surprised that the solution could be expressed in terms of a family of congruent integers. This also suggests that Dan's understanding of ROM was highly procedural.

In the interview, Barbara immediately lamented the fact that she did not really understand why the ROM procedure worked, nor why one has to go back through and unravel it to get the solution:

> I understand why we're doing that, reducing it down, but when we start introducing other variables and you know, keep trying to reduce it, reduce it, and then probably where I get lost in all the thinking is when we go back to unraveling it. I'm trying to figure out like why that's important to solve it. You know to me, it seems like once we get down here, that would seem like a solution but it's not because you have to go back and do that so, that's what's the mystery to me.

When asked how she knew when to stop the reduction process, she said, "When I have a coefficient of 1 on the left side.... There's been some spots in some of the examples in the book where there will be something like this ... with like a higher coefficient like a 6 or something, and then automatically it's gone to what m is by itself. But like I can't figure out how to jump from here to m without going through all the middle steps."

Solving Linear Congruences. All of the participants except Andrew used ROM to solve the linear congruences on the exam, despite the fact that they were not taught to solve congruences using this method. Dr. Thomas always used a guess and check method to solve congruences in class, and only used ROM when solving linear Diophantine equations. The participants appropriated this method to solve congruences, and generally were not aware of or rejected outright the possibility of using the guess and check strategy. As a reminder, the congruences discussed in this section were from the exam covering congruence: (a) $5x \equiv 1 \pmod{11}$ and (b) $6x \equiv 3 \pmod{15}$.

Fran remarked that solving these two congruences was much easier than the linear Diophantine problems, because these were "already set up for reduction of moduli." This comment is striking—she is saying that this is the *same* type of problem as the Diophantine equation. When asked how else she could have approached solving these congruences, she said that she could have rewritten them [like $5x - 11y = 1$] and used the Euclidean algorithm. I pressed Fran to think of other ways to solve these congruences, and finally suggested the guess and check strategy. Despite the fact that Dr. Thomas had used this strategy in class exclusively, Fran responded that she would probably never have used this strategy. She said that it would "take forever to solve the problem that way, because there's just way too many possibilities of xs and ys that you could try before you would actually come up with an answer that would work."

Eva said that in this problem, she was "looking for a set of numbers [for x] that if you plug them in then it would be true that if you divide it by 11, you would get a remainder of 1." Her work for this problem showed an incredibly long process, considering that one could easily solve this problem using a guess and check strategy. Eva realized this spontaneously in the interview, saying that all she really needed to do was just try 0 through 10 for x, and then check to see if this produced a remainder of 1.

Dan said that he was looking for "5 times some number that is gonna give you, if divided by 11, a remainder of 1, something that would be in the 1 mod 11 family." He said that for (a), his first thought was just to try each of the 11 possible values for x to see which would work, but then decided to use ROM instead because this seemed to be the "more theoretical way" of doing the problem.

When asked what she was looking for in (a), Barbara said, "We're looking for the set of numbers such that when you divide them by 11, you're left with a remainder of 1." When she described her solution, however, she explained, "I can rewrite (a) like $5x - 1 = 11y$, and then it would have been basically the same problem that [the Diophantine equation] was." I asked if she thought of congruences as being similar to equations, and she said, "I actually look at them in terms of an equation. Like when I look at that, I'm thinking to myself, $5x$ equals 1 plus $11y$."

Most of the participants seemed reluctant to view congruences as analogous to equations. They did not frequently operate on both sides of a congruence, and did not view solving linear congruences as similar to solving linear equations at all. Andrew was the only participant who appeared to work with congruences as if they were analogous to equations. He solved congruences by substituting more convenient representatives on the right of the congruence and then cancelled numbers on both sides. Although he did not know when this process actually was allowable and when it was not, he was at least viewing congruences relationally in this context. He described the process using words that were reminiscent of solving equations: "You're sort of like, I don't know if you want to say balancing, but you're sort of working on both sides of the congruence."

Reasoning About Congruences Using Equations. In general, the participants were reluctant to think of congruences as similar to equations, and avoided operating on both sides of a congruence by rewriting the congruence as an equation and operating with this equation. At the time of the second interview, Fran seemed to believe that congruences were just different notations for equations in two variables. She had been struggling to understand how to operate within the mod n world, and although she would typically interpret a congruence in terms of division and remainder, when working with congruences, the first thing she wanted to do was to rewrite them as equations. This new representation for congruence became a means by which Fran could *avoid* thinking about the mod n world. When asked if she viewed working with congruences as similar to working with equations, and she replied, "To me it's not an equation like that, but I know I can convert it into an equation. But I don't look at that as an equation."

Eva used ROM to solve problem (b) [$6x \equiv 3 \pmod{15}$], first reducing the coefficients modulo 6, and then reducing $3a = 3 \pmod 6$ to $a \equiv -1 \pmod 2$ by dividing through by 3. When I asked why one can divide through a congruence like this, she struggled to explain using ideas related to congruence, but then decided to explain by rewriting the congruence as an equation.

> You can just go ahead and take out that it's 3 times 11 is equal to 3 times 1 plus $2x$, and then you can just go ahead and cancel the 3 and then you get 11 is equal to 1 plus $2x$, which is the same thing as if I would have it here, 11 is congruent to negative 1 mod 2. I can't explain it in a mod 6 way because it's a little more high level of thinking, but if I could change that back to an equation, which is the way we've been doing it and that's fine and that would work.

Unlike Fran, Eva seemed to use the equation representation as a tool for working with congruences, rather than rewriting congruences as equations to avoid thinking about mod world.

Although he seemed to have a solid understanding of ROM, Chris was quick to admit that there were things he did not understand about working with congruences. For example, before he divided through a congruence by a number, he said that he had to stop and prove it to himself by rewriting the congruence as an equation, even though he knew that there was a theorem that said that he could divide through a congruence by the GCD.

> Every time I think that through. I can almost consciously remember thinking, what if I divide everything by 3, is that correct? Can I do that? And so then, I went back to this, set it up this way [as an equation in $\mathbb{Z}$], and divided by 3 and said, yeah. I understood the idea of dividing by 3 by testing out the theory, even though it is a theorem we have. I guess that's my defense, my way of stopping from making that mistake is to consciously think about it each time that subtracting 1 from both sides doesn't change. It's still the same thing. No, that's not true ... It doesn't change the relationship of this side to this side.

The use of the word *relationship* is interesting in this context, because he generally appeared to view congruence operationally when solving problems.

Dan and Barbara both seemed comfortable rewriting congruences as equations, and used this strategy to explain why one can divide through a congruence by a number. Barbara rewrote $6m \equiv 9 \pmod{21}$ as $6m - 21n = 9$, and then noted that both sides were divisible by 3. In general, Barbara seemed unaffected by the abstraction involved in dealing with congruence. Whereas the other participants struggled to make sense of the mod world, Barbara showed a tendency to convert congruences to equations even in the first interview. As a result, she was very proficient at manipulating symbols without much thought. She achieved the highest score in the class on the second exam, although there were several gaps in her understanding of congruence.

Dan noted that working in the mod world was uncomfortable, and stated that it was much easier to work with congruences after converting them to equations.

> People are uncomfortable with congruence arithmetic and they look at this and they say, I don't really understand the rules of congruence arithmetic. But if you put it straight out in normal equation setting, you can see that obviously, the 3 can [be factored] out and so, that's not a problem.

The other students occasionally performed an operation on both sides of the congruence, such as multiplying both sides by the same number, but they were more likely to remove a number from a congruence, perform a string of operations on it, such as dividing it by a number and keeping the remainder, and then writing a new congruence showing the completion of this process.

DISCUSSION

By the second interview, most of the participants showed a tendency to avoid working in what they called the mod world. Some were explicit about it, expressing a preference for thinking about congruences as equations in $\mathbb{Z}$. The most interesting instance of this avoidance was seen on the exam, when all of the participants except Andrew solved the linear congruences using reduction of moduli, despite the fact that Dr. Thomas never solved congruences this way in class. Why did the students choose to use reduction of moduli to solve linear congruences?

If the congruences in question did not have such small moduli, the answer to this question might be more straightforward. For example, if the students were presented with a congruences such as $5x \equiv 1 \pmod{37}$, it might be understandable that they would be reluctant to use the brute force strategy introduced in class. However, this was not the case for the congruences they were to solve on the exam, such as $5x \equiv 1 \pmod{11}$. There are only 11 possibilities, and 0 and 1 can be disregarded quickly. If the student notices that $5 \times -2 = -10$, and then recognizes that $1 \equiv -10 \pmod{11}$, the solution can be found even more quickly. In fact, Dr. Thomas solved many congruences in class by trying various small positive and negative values for x, so this was a strategy the students had seen.

It is also interesting to note that the students generally did not seem to understand the value of using congruence of integers to solve problems. Consider the way that the students were introduced to congruence: First, they were introduced to linear Diophantine equations and to some unwieldy methods for solving them. Then, they were introduced to linear congruences and given a method for solving these. Finally, Dr. Thomas directed the class's attention back to the linear Diophantine equations, stating that they now would learn a method of solving these using congruence of integers, which would give all solutions to these equations.

The students did use this strategy to solve linear Diophantine equations. However, they also used this strategy to solve congruences. In effect, they reversed the process they had been taught: They started with a congruence, rewrote it as an equation in $\mathbb{Z}$, and then solved this equation using ROM, which is significantly more complicated than trying a handful of values for x. The power of using congruence was lost on them, and they appropriated a strategy that allowed them to avoid solving the congruence in the mod world.

Why did this happen? Clearly the students were struggling with the concept of congruence; perhaps it is human nature to use a more complicated strategy in a familiar context to avoid this struggle. Working with the equation representation was more familiar than working with a congruence statement. Several of the participants, including Barbara and Chris, admitted that they preferred to convert congruences to equations immediately. Chris even stated that he frequently performed this conversion first, without considering whether or not it would be helpful in the situation.

The fact that congruence is an equivalence relation was not emphasized in class, although it was in the text. All of the students except Fran had taken the department's "transition" course (Introduction to Mathematical Reasoning) in which they had been introduced to equivalence relations. Because the students were not specifically exposed to the fact that congruence is an equivalence relation, and may not have read this in the text, their introduction to congruence was quite similar to the way children are introduced to the concept of equality. They were given several equivalent definitions for congruence in class, but then appeared to develop their conceptions of congruence primarily from working with examples. Their tendency to interpret congruence operationally could be a product of many factors. For example, the participants tended to use the division–remainder representation for congruence as a definition, disregarding the fact that it is not equivalent to any definition of congruence.

In addition, most of the participants believed that the position of an integer in a congruence determined the possible range of values that it could take, as described in the previous chapter. For some of the students, congruences were initially interpreted as statements about division and remainder, with the number on the left as the dividend and the number on the right the remainder. By the second interview, most viewed congruences as statements about integers being transformed into numbers existing in the mod n world. In most cases, linear congruences to be solved were written with the variable on the left side of the congruence, just as linear equations are typically written in algebra. When the congruence was missing a value on the right, the intention was that the students find the standard representative for a generally large number on the left. Similarly, children typically see arithmetic equalities written as $a + b = c$, so that the operation of addition can be thought of as being performed left to right, with the answer, c, written on the right. Children rarely see statements like $? = 3 + 5$. Algebra students generally solve equations in which the variable is on the left as well, such as $23x + 14 = 12$. High school mathematics teachers frequently experience the scenario in which students do not know how to solve equations in which like terms must first be collected: $14x + 5 = 7x - 8$. If the students have an operational view of equality, as mentioned earlier in the chapter, the process of solving equations by performing operations to both sides of an equation is difficult to understand.

It is interesting to note there was little apparent transfer between the participants' prior knowledge of working with algebraic equations to working with analogously structured linear congruences. For the participants, these were two different types of problems. In fact, as noted earlier, they tended to view the process of solving linear Diophantine equations and linear congruences as one and the same. There was no recognition that one of these was an equation of two variables in $\mathbb{Z}$, and the other was an equation of one variable in the "world mod n."

Upon reflection, I think one of the conceptual difficulties inherent in learning the concept of congruence is the fact that the numbers in a congruence statement can represent several different mathematical objects. From one perspective, we can think about congruence explicitly as an equivalence relation defined on $\mathbb{Z}$; then there are infinitely many integers that one can substitute on either side of a given congruence. From this perspective, these integers are representing their respective congruence classes, and so there would be infinitely many solutions to a problem like $5x \equiv 1 \pmod{11}$, all of the form $9 + 11k$, for $k \in \mathbb{Z}$. On the other hand, we can also think of congruences as equations in $\mathbb{Z}/n\mathbb{Z}$, and regard the equivalence classes as the objects we can manipulate. In this case, there would only be one solution to the above congruence; the equivalence class denoted by 9. Another possible interpretation is to think of the set $\{0, 1, 2, \ldots n - 1\}$ as a finite ring of integers, with addition and multiplication defined modulo n. In this case, the congruence would also have only one solution.

All of these interpretations are encapsulated in this single representation, and mathematicians are able to work with this ambiguity with little difficulty. In fact, many mathematicians might refuse to distinguish between these interpretations at all. However, for students who are just beginning to make sense of this concept, these distinctions are real and confusing. It should not be surprising that the six participants in this study had difficulty interpreting congruences, or that they made sense of congruence by combining the prior interpretations into a single conception.

A confounding factor that likely influenced the students' view of position was a lack of distinction between the operation of reducing integers modulo n and congruence. Because I chose to focus on the relation of congruence, I did not systematically collect data related to their understanding of this operation. This is a question for future research.

It is certainly true that in practice, most of the congruences the students wrote or saw written had the property that the integer to the right of the $\equiv$ symbol was smaller than the modulus. If the congruence was an equation to be solved, the variable would generally be located on the left side of the congruence. In elementary algebra, we almost always write equations this way. Children often see arithmetic identities written as $2 + 3 = 5$, not $5 = 2 + 3$. Similarly, perhaps points lost for giving the unreduced form of an answer on the exam may have contributed to the students' beliefs that the integer on the left side of the congruence has to be reduced. The role of convention in contributing to students' conceptions of congruence is another aspect that was not systematically explored, although I did obtain some data concerning this topic.

Each of the participants seemed to believe at some point that the position of an integer in a congruence was significant; they either disregarded or were unaware of the symmetric nature of the congruence relation. At the time of the first interview, Andrew, Dan, Eva, and Fran all believed relative position was important, whereas Chris and Barbara did not. Three weeks later, everyone except Andrew believed position was important, and Andrew had not completely resolved the issue. The case of Chris is particularly interesting—why did he think about congruences primarily as equivalence relations at the time of the first interview, and not at the time of the second? The fact that this shift occurred is significant, because all of the participants seemed to have otherwise deepened their understandings of congruence during that time.

Concluding Remarks

Congruence of integers is a concept that seems to lie in the borderlands of elementary and advanced mathematical thinking. It is a topic that can, in a limited fashion, be approached from a very elementary perspective. After all, children learn how to do clock arithmetic in elementary school. However, the student must make the shift to thinking about congruence from an advanced perspective in order to develop a rich understanding of congruence as an equivalence relation. It seems that advanced mathematical thinking may be necessary in order for the student to make sense of even simple statements of congruence.

The goal of the course was not to prepare students for further study in algebra or number theory. Rather, the intent of Dr. Thomas was that the students see congruence arithmetic as an interesting way of looking at numbers, as well as a powerful tool for solving problems. However, even studying congruence for these humble purposes requires students to use advanced mathematical thinking. The department's transition course was a prerequisite for the number theory course studied, but this policy was not enforced. This study implies that an awareness of the types of difficulties that may be encountered by students who have not yet made the transition to advanced mathematical thinking might be helpful for instructors of number theory at this level.

Much of the difficulty the students had seemed to be rooted in the vagueness of the notation of congruence. By the second interview, the students had many ideas about congruence that were disorganized and disconnected. One issue with which the students seemed to wrestle was that different interpretations of congruence seemed to be used in various contexts. In one type of problem, congruences would be rewritten as equations, and in another, the divisibility definition would be used. For several of the participants, there were no connections between these representations, and their understandings of congruence seemed to become more fragmented with time. The roles of these different interpretations of congruence were not dealt with in class by Dr. Thomas, and I doubt that most instructors of such a course would discuss them. I do not mean to suggest instructors are not aware of the difficulties students have with this concept, but I think the different interpretations of the concept are so well connected in the minds of most mathematicians that it would seem strange to pick the concept apart in class and discuss the details with the students. This issue of what knowledge of a concept is needed for teaching is an important one in mathematics education. In order to effectively teach a concept, the instructor should have what Ball, Bass, Hill, and Hoover (2001) called "unpacked" mathematical knowledge. The power of mathematics lies in its compactness, and mathematicians are incredibly proficient at thinking abstractly. Once one understands how to think so abstractly, there is great power in doing so.

Recommendations

A question raised by this research is whether it would be better to present congruence in a highly formalized way than to use the informal approach taken by Dr. Thomas. Although I have no specific evidence, I doubt the results would have been qualitatively different. The idea of building on students' prior understanding of division with remainder and divisibility is pedagogically sensible and aligns with most theories of mathematical learning. Moreover, it also reflects the historical development of the concept. Similarly, no one advocates that we introduce children to the concept of function by giving them the formal set theoretic definition. Children's understanding of function is developed by building on their natural understanding of relationships between quantities; profit from a lemonade stand is a function of the number of cups of lemonade sold, for example. Once this understanding is in place, the notion of function can be refined to include univalence, and can be extended to arbitrary sets.

One recommendation I would make is this: The fact that congruence is an equivalence relation should be emphasized by reinforcing the fundamental idea that two congruent integers are considered to be the same in a certain sense. Without this notion of equivalence, the operational interpretation of congruence seems almost a natural interpretation. Just as elementary teachers should avoid always writing equalities with numbers and arithmetic operations on the left equal to a single number on the right, perhaps we should be conscious of how we write congruences.

The distinction between the process of reducing mod n and using congruence to denote that two integers are equivalent should be made clear to students. This is similar to the distinction between the operational and relational interpretations of equality. We want children to understand that the equality symbol can be used in a variety of ways; in both $12 + 5 = ?$ and $12 + 5 = ? + 11$, there is an answer to be found, but in both cases, = indicates that the quantities on both sides of the symbol are the same. We do use = as a "find the answer" symbol, but there is a deeper meaning there as well. Similarly, we do write congruences that indicate a reduction has occurred, such as $25 \equiv 7 \pmod{15}$. However, this congruence could also be written as $7 \equiv 25 \pmod{15}$, and we could have chosen to represent 25 by –8 as well.

In teaching students to solve linear congruences, it could be beneficial to treat these congruences as analogous to equations from the beginning. Kieran (1999) guided middle school algebra students to think of arithmetic identities relationally by asking them to build identities such as $(5 + 2) \cdot 3 = 2 \cdot 10 + 1$. Eventually, these were replaced by equations by making one of the numbers a variable, so that the students would try to figure out what number the x could be in $3 \cdot 7 + 3 = x - 1$. The process of building the arithmetic identities led the students in this teaching experiment to view equality relationally; the process of learning to solve equations was then based on this foundation. As noted earlier, methods typically used to introduce children to solving equations tend to reemphasize an operational interpretation of equality, such as using a guess-and-check strategy to solve $3x - 4 = 17$. It is not that this type of strategy is not valuable, but to a child with an operational understanding of equality, solving this equation becomes a process of finding a way to make 17, instead of trying to find an x such that both sides of the equation are the same. When the equations become more complicated, the student has no recourse but to memorize a complex and fairly meaningless procedure in order to solve the equation. For students with a relational view of equality, solving such equations is not a matter of remembering how to do the problem, but a solution process that makes sense.

For the students in this study, solving a congruence like $2x \equiv 1 \pmod 3$ was easy; they could quickly check that substituting 2 for x and dividing the product by 3 gives a remainder of 1. However, solving $5x \equiv 1 \pmod{11}$ appeared to be less straightforward. I suggest that giving students opportunities to learn to solve congruences by performing operations on both sides, similar to the equivalent equations method for solving equations in algebra, would not only reinforce a relational view of congruence, but would make the students more comfortable in the mod world. For example, the students could approach solving the earlier congruence by thinking, "I need to multiply both sides of the congruence by the inverse of 5." The concept of multiplicative inverse could be introduced early, perhaps when the students are learning about the Euclidean algorithm. The students could build multiplication tables for certain mod n worlds and examine patterns that appear. One could imagine that this type of activity would be very valuable for future middle and high school teachers. The students should be able to discover for themselves which integers are invertible modulo n and which are not, and they should be able to prove this fact as well. Once the existence of modular inverses has been established, students can write simple calculator programs to compute them for any value of n. Then, solving a congruence such as $5x \equiv 1 \pmod{11}$ is simply a matter of multiplying both sides of the congruence by 9, which is the inverse of 5.

I remain convinced that it could be valuable in a number theory course (or in a capstone course) to give prospective teachers experiences in solving congruences using methods that are analogous to those used in algebra. Coupled with discussions of the similarities and differences between working in the integers or rationals and working in finite number systems, these experiences may serve to deepen their understandings of the processes of the mathematics they will be teaching. Postsecondary mathematics faculty often remark that their freshmen do not understand basic algebra, and wonder what high school teachers are doing in their classrooms. I would suggest that for most teachers, their university education did little to prepare them to teach algebra effectively. Many teachers in the field may deepen their understanding of algebra on their own, by attending workshops, conferences, and continuing to study mathematics. Others choose to focus instead on one of the myriad other problems that they encounter daily in their classrooms, such as working with students with special needs, finding ways to better teach other mathematical subjects, and struggling to accommodate the needs of an incredibly diverse population. I firmly believe it is the responsibility of university mathematics faculty to prepare and support these teachers, if for no other reason than that they in return will provide us with more prepared students in the future.

APPENDIX: SOLVING CONGRUENCES

Solving Linear Congruences

- $5x \equiv 1 \pmod{11}$: Dr. Thomas expected the students to solve this congruence using a "guess-and-check" strategy. Since there are only 11 classes of integers to consider, one only has to try a representative from each. Once a representative that satisfies the equation is found, all integers in that congruence class are known to be solutions as well. In this case, –2 is a solution, so all integers that are congruent to –2 modulo 11 are solutions to the congruence. As with linear equations in R, this is the only solution.
- $6x \equiv 3 \pmod{15}$: This case is more complicated in that the students were expected to notice that a factor of 3 can be removed from the congruence to obtain $2x \equiv 1 \pmod 5$. They had learned the following theorem in class: If $ax \equiv b \pmod n$, $\gcd(a,n) = d$, and $d \mid b$, then $x \equiv (\mathrm{mod}\ )$. The resulting congruence could then be solved by trial and error, as described above.
- $19^{19} \equiv$ __ (mod 7): There were several ways the students could have solved this congruence, including brute force computation on their calculators. Dr. Thomas's intention was that they employ Fermat's "little" theorem: If a and p are relatively prime, then $a^{p-1} \equiv 1 \pmod p$. Notice that . Fermat's "little" theorem implies that $19^6 \equiv 1 \pmod 7$, and so by substitution, we find that $19^{19} \equiv 1 \cdot 19 \pmod 7$. Thus, any integer that is in the congruence class of 19 modulo 7 satisfies this congruence. The smallest such integer is 1.

Reduction of Moduli

Reduction of moduli (ROM) was introduced in the course as a technique for solving linear Diophantine equations (equation in two or more variables in $\mathbb{Z}$ of the form $ax + by = c$). The basic premise of ROM modulo is that one views the Diophantine equation modulo one of its coefficients, so that the original problem can be reduced to that of solving a linear congruence. Once the congruence is solved, this solution can be transferred back to $\mathbb{Z}$ to obtain the set of all solutions to the equation.

The students used ROM to solve equations with fairly large coefficients, and only knew how to solve congruences by trying all possibilities, so the ROM process was actually a series of iterations of the above process. For example, a Diophantine equation to be solved on the exam was $75x + 27y = 12$. First, we view the equation as a congruence, modulo 27, and then reduce the coefficients:

$$75x + 27y \equiv 12 \pmod{27}$$
$$21x \equiv 12 \pmod{27}$$

Because this congruence implies that $27 \mid 21x - 12$, or $(3 \cdot 9) \mid 3 \cdot (7x - 4)$, we can write the following congruence:

$$7x \equiv 4 \pmod{9}$$

At this point, we could solve the congruence by trying the seven possibilities for x, or we could transfer the congruence back to $\mathbb{Z}$, and get an equation related to the original, $7x = 4 + 9k$. Now we can repeat this process by viewing this equation as a congruence, modulo 7:

$$7x \equiv 4 + 9k \pmod{7}$$

We can reduce the coefficients modulo 7, and cancel a 2 from both sides because 7 is prime (and $\mathbb{Z}/7\mathbb{Z}$ is a field):

$$-4 \equiv 2k \pmod{7}$$
$$-2 \equiv k \pmod{7}$$

At this point, we can write an expression for $k = -2 + 7n$ and substitute this into the equation $7x = 4 + 9k$ to get $x = -2 + 9n$, and then substitute this into the original equation to get $y = 6 - 25n$. Now we can generate all possible solutions for the equation by substituting in values for n; for example, $n = 0$ gives the solution $(-2, 6)$.

REFERENCES

Ball, D. L., Bass, H., Hill, H., & Hoover, M. (2001). *Understanding teaching as mathematical problem solving.* Retrieved April 1, 2002, from http://www-personal.umich.edu/~dball/BallSelectedPresentations.html

Ball, D. L., & McDiarmid, G. W. (1990). The subject-matter preparation of teachers. In W. R. Houston, M. Haberman, & J. P. Sikula (Eds.), *Handbook of research on teacher education* (pp. 437–449). New York: Macmillan.

Behr, M., Erlwanger, S., & Nichols, E. (1980). How children view the equals sign. *Mathematics Teaching, 92,* 13–15.

Bernard, J. E., & Cohen, M. P. (1988) An integration of equation-solving methods into a developmental learning sequence. In A. F. Coxford (Ed.), *The Ideas of algebra, K–12* (pp. 97–111). Reston, VA: National Council of Teachers of Mathematics.

Booth, L. R. (1988). Children's difficulties in beginning algebra. In A. F. Coxford (Ed.), *The ideas of algebra, K–12* (pp. 20–32). Reston, VA: National Council of Teachers of Mathematics.

Brown, A., Thomas, K., & Tolias, G. (2002). Conceptions of divisibility: Success and understanding. In S. Campbell & R. Zazkis (Eds.), *Learning and teaching number theory: Research in cognition and instruction* (pp. 41–82). Westport, CT: Ablex.

Campbell, S. (2002). Coming to terms with division. In S. Campbell & R. Zazkis (Eds.), *Learning and teaching number theory: Research in cognition and instruction* (pp. 15–40). Westport, CT: Ablex.

Clark, J. M., DeVries, D. J., Hemenway, C., St. John, D., Tolias, G., & Vakil, R. (1997). Introduction. *The Journal of Mathematical Behavior, 16*(3), 181–185.

Conference Board of the Mathematical Sciences (CBMS). (2001). *The mathematical education of teachers.* Washington, DC: American Mathematical Society.

Cuoco, A. A. (2001). Mathematics for teaching. *Notices of the American Mathematical Society, 48*(2), 168–174.

Dubinsky, E., Dautermann, J., Leron, U., & Zazkis, R. (1994). On learning fundamental concepts of group theory. *Educational Studies in Mathematics, 27,* 267–305.

Fennema, E., & Franke, M. L. (1992). Teachers' knowledge and its impact. In D. A. Grouws (Ed.), *Handbook of research on mathematics learning and teaching* (pp. 147–164). New York. MacMillan.

Ferrari, P. L. (2002). Understanding elementary number theory at the undergraduate level: A semiotic approach. In S. Campbell & R. Zazkis (Eds.), *Learning and teaching number theory: Research in cognition and instruction* (pp. 97–116). Westport, CT: Ablex.

Freudenthal, H. (1983). *Didactical phenomenology of mathematical structures.* Boston, MA: Kluwer.

Hazzan, O. (1999). Reducing abstraction level when learning abstract algebra concepts. *Educational Studies in Mathematics, 40,* 71–90.

Herscovics, N., & Kieran, C. (1999). Constructing meaning for the concept of equation. In B. Moses (Ed.), *Algebraic thinking, grades K–12* (pp. 181–193). Reston, VA: National Council of Teachers of Mathematics.

Kieran, C. (1981). Concepts associated with the equality symbol. *Educational Studies in Mathematics, 12,* 317–326.

Kieran, C. (1988). Two different approaches among algebra learners. In A. F. Coxford (Ed.), *The ideas of algebra, K–12* (pp. 91–96). Reston, VA: National Council of Teachers of Mathematics.

Kieran, C. (1999). The learning and teaching of school algebra. In B. Moses (Ed.), *Algebraic thinking, grades K–12* (pp. 341–361). Reston, VA: National Council of Teachers of Mathematics.

Ma, L. (1999). *Knowing and teaching elementary mathematics*. Mahwah, NJ: Lawrence Erlbaum Associates.

National Council of Teachers of Mathematics (NCTM). (2001). *Navigating through algebra in grades 9-12*. Reston, VA: Author.

O'Connor, J. J., & Robertson, E. F. (2002). *Abu Ali al-Hasan ibn al-Haytham*. Retrieved February 19, 2002, from http://turnbull.mcs.stand.ac.uk/history/Mathematicians/Al-Haytham.html

Ore, O. (1948). *Number theory and its history.* New York: McGraw-Hill.

Rowland, T. (2002). Generic proofs in number theory. In S. R. Campbell & R. Zazkis (Eds.), *Learning and teaching number theory: Research in cognition and instruction* (pp. 157–184). Westport, CT: Ablex.

Saenz-Ludlow, A., & Walgamuth, C. (1998). Third graders' interpretations of equality and the equal symbol. *Educational Studies in Mathematics, 35,* 153–187.

Selden, A., & Selden, J. (2002). Reflections on mathematics education research questions in elementary number theory. In S. Campbell & R. Zazkis (Eds.), *Learning and teaching number theory: Research in cognition and instruction* (pp. 213–230). Westport, CT: Ablex.

Smith, J. C. (2002). *An investigation of undergraduates' understanding of congruence of integers*. Doctoral dissertation, University of Arizona. Dissertation Abstracts International.

Strauss, A., & Corbin, J. (1998). *Basics of qualitative research: Techniques and procedures for developing grounded theory* (2nd ed.). Thousand Oaks, CA: Sage.

Usiskin, Z. (1988). Conceptions of school algebra and uses of variables. In A. F. Coxford (Ed.), *The ideas of algebra, K–12* (pp. 8–19). Reston, VA: National Council of Teachers of Mathematics.

Wagner, S. (1977). *Conservation of equation, conservation of function, and their relationship to formal operational thinking*. Unpublished dissertation, New York University.

Wagner, S., & Parker, S. (1999). Advancing algebra. In B. Moses (Ed.), *Algebraic thinking, grades K–12* (pp. 328–340). Reston, VA: National Council of Teachers of Mathematics.

Zazkis, R. (1998). Odds and ends of odds and evens: An inquiry into students' understanding of even and odd numbers. *Educational Studies in Mathematics, 36,* 73–89.

Zazkis, R. (1999). Challenging basic assumptions: Mathematical experiences for pre-service teachers. *International Journal of Mathematical Education in Science and Technology, 30*(5), 631–650.

Zazkis, R., & Campbell, S. (1996). Divisibility and multiplicative structure of natural numbers: Pre-service teachers' understanding. *Journal for Research in Mathematics Education, 27*(5), 540–563.

Author Index

Subject Index

I

J

K

L

M

N

O

P